SpringerBriefs in Applied Sciences and Technology

PoliMI SpringerBriefs

Springer, in cooperation with Politecnico di Milano, publishes the PoliMI Springer-Briefs, concise summaries of cutting-edge research and practical applications across a wide spectrum of fields. Featuring compact volumes of 50 to 125 (150 as a maximum) pages, the series covers a range of contents from professional to academic in the following research areas carried out at Politecnico:

- Aerospace Engineering
- Bioengineering
- Electrical Engineering
- Energy and Nuclear Science and Technology
- Environmental and Infrastructure Engineering
- Industrial Chemistry and Chemical Engineering
- Information Technology
- Management, Economics and Industrial Engineering
- Materials Engineering
- Mathematical Models and Methods in Engineering
- Mechanical Engineering
- Structural Seismic and Geotechnical Engineering
- Built Environment and Construction Engineering
- Physics
- Design and Technologies
- Urban Planning, Design, and Policy

More information about this subseries at https://link.springer.com/bookseries/11159
http://www.polimi.it

Manuela Antonelli · Gabriele Della Vecchia
Editors

Civil and Environmental Engineering for the Sustainable Development Goals

Emerging Issues

POLITECNICO
MILANO 1863

Springer

Editors
Manuela Antonelli
Department of Civil and Environmental
Engineering (DICA)
Politecnico di Milano
Milan, Italy

Gabriele Della Vecchia
Department of Civil and Environmental
Engineering (DICA)
Politecnico di Milano
Milan, Italy

ISSN 2191-530X   ISSN 2191-5318 (electronic)
SpringerBriefs in Applied Sciences and Technology
ISSN 2282-2577   ISSN 2282-2585 (electronic)
PoliMI SpringerBriefs
ISBN 978-3-030-99592-8   ISBN 978-3-030-99593-5 (eBook)
https://doi.org/10.1007/978-3-030-99593-5

This Springer imprint is published by the registered company Springer Nature Switzerland AG
The registered company address is: Gewerbestrasse 11, 6330 Cham, Switzerland

Preface

The fast evolution of our society often falls short in properly taking into consideration its relationship with the environment, which is not only the primary source of any resource and the sink of all the wastes we generate throughout our activities, but also the cause of most of the loading and constraints applied to structures and infrastructures. The lack of proper consideration of the relationship between the needs of the society and the environment may lead to strong disequilibria, generating a large amount of threats for robust, resilient and continuous development. The scarcity of resources thus threatens global economic and social stability.

In this perspective, the sustainable development goals (SDGs) set by the United Nations represent the criteria to revise our development model, towards the ability to conjugate different needs to build a safe relation between anthropic activities and the environment. According to the United Nations Development Program, SDGs (adopted by the United Nations in 2015) represent a *universal call to action to end poverty, protect the planet, and ensure that by 2030 all people enjoy peace and prosperity*. The 17 SDGs recognize that actions in one area will affect outcomes in other areas and that development must balance social, economic and environmental sustainability.

It is evident that Civil and Environmental Engineering should play a relevant role in providing methods, approaches, risk and impact assessments, as well as technologies, to fulfill the SDGs. Research in these fields may in fact provide technical knowledge and tools to support decision-makers and technicians in (i) planning mitigation and adaptation actions to climate change, extreme weather events, earthquakes, drought, flooding and other natural disasters; (ii) designing efficient strategies for resources exploitation; (iii) increasing the safety of structures and infrastructures under exceptional loadings and against the deterioration due to their lifecycle; (iv) adopting a holistic risk management approach to reduce environment deterioration.

This volume collects some emerging issues in Environmental and Civil Engineering, coming from selected Ph.D. Theses discussed at Politecnico di Milano in 2021. All the contributions belong to either the Doctoral Programs in Environmental and Infrastructure Engineering (IAI) or Structural, Seismic and Geotechnical Engineering (ISSG). The advanced innovative insights provided are discussed with reference to the relevant SDGs. In particular, 6 out of 17 SDGs are addressed in this volume:

- SDG 2 (Zero Hunger),
- SDG 6 (Clean Water and Sanitation),
- SDG7 (Affordable and Clean Energy),
- SDG9 (Industry, Innovation and Infrastructure),
- SDG11 (Sustainable Cities and Communities),
- SDG12 (Responsible Consumption and Production).

SDG 2 (Zero Hunger), SDG 6 (Clean Water and Sanitation), SDG7 (Affordable and Clean Energy), SDG9 (Industry, Innovation and Infrastructure), SDG11 (Sustainable Cities and Communities) and SDG12 (Responsible Consumption and Production). The topics addressed in the volume cover developments and innovation in the fields of

- as for IAI Ph.D., drinking water treatment and management (Chap. 1), organic waste recovery for energy production (Chap. 2), flood damage assessment (Chap. 3) and remote sensing for environmental monitoring (Chap. 4);
- as for ISSG Ph.D., energy harvesting via metamaterials (Chap. 5), stability of masonry structures of historical interest (Chap. 6), design of sustainable piled embankments reinforced with geosynthetics (Chap. 7) and monitoring of unstable rock slopes with microseismic methods (Chap. 8)

(i) as for IAI Ph.D., drinking water treatment and management (Chap. 1), organic waste recovery for energy production (Chap. 2), flood damage assessment (Chap. 3) and remote sensing for environmental monitoring (Chap. 4); (ii) as for ISSG Ph.D., energy harvesting via metamaterials (Chap. 5), stability of masonry structures of historical interest (Chap. 6), design of sustainable piled embankments reinforced with geosynthetics (Chap. 7) and monitoring of unstable rock slopes with microseismic methods (Chap. 8).

The addressed topics apply to the everyday life of citizens, even if to a non-expert they could be far from that. As a matter of fact, the developed research work provides insights and supporting tools to technicians, practitioners, decision-makers, regulatory and control authorities, each one in his specific field of expertise. This collection has the ambition to ease the transfer of knowledge and good practices in order to contribute to switch the current paradigm about the relationship between human beings and the environment: our society is called upon to take preventive actions rather than mitigation and remediation actions as occurred so far, so we only realized that we had generated a problem when that problem was so big that it could no longer be ignored.

The coexistence of different research areas proves the central role played by Civil and Environmental Engineering in shaping a sustainable future, promoting the collaboration between young researchers and established open and dynamic research groups. In conclusion, the aim of this collection is to provide a cutting-edge overview of the most recent research trends in the Department of Civil and Environmental

Engineering (DICA) of Politecnico di Milano, in an easy-to-read format to present the main results even to non-specialists in the specific field.

Researchers too often want to publish their work on high impact factor journals for experts only, considered a great achievement, but the question we are called to answer is what happens next? To assure a concrete impact on our society, it is important not only to publish open access, to permit broad accessibility to information, but also to explain concepts in a comprehensible way and to make them accessible also to non-specialists. This means to simplify the language, while conveying accurate and not misleading messages, and to adopt complementary forms of communications, such as graphical abstracts and short summaries. Then, we decided to accept this challenge to improve our scientific communication.

Hopefully, this volume will also provide a boost for having sustainable development as the guiding star for modern, impacting and inclusive research in the field of Civil and Environmental Engineering.

Milan, Italy   Manuela Antonelli
December 2021   Gabriele Della Vecchia

Contents

A Risk-Based Approach for Contaminants of Emerging Concern in Drinking Water Production and Distribution Chain

Beatrice Cantoni ⓘ

Abstract Provision of safe drinking water (DW) is one of the major requisites for human health, related to four Sustainable Development Goals (SDGs) of the United Nation 2030 Agenda: SDGs 3 (Good health), 6 (Clean water and sanitation), 11 (Sustainable cities) and 12 (Responsible production and consumption). However, this is hindered by the presence, especially in highly-anthropized contexts, of contaminants of emerging concern (CECs) in DW, that may pose a risk for human health. The present study aims at developing a holistic framework to support both (i) decision-makers for CECs prioritization in DW regulation and (ii) water utilities for the selection of appropriate monitoring and treatment interventions for the optimization of DW supply system. In detail, a quantitative chemical risk assessment (QCRA), including uncertainties related to both exposure and hazard assessments, was developed. Then, it was combined with testing and modeling of CECs fate in treatment processes and in distribution network, obtaining a robust tool to achieve the above-mentioned SDGs.

Graphical Abstract

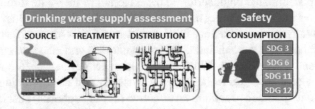

Keywords Activated carbon · Contaminants of emerging concern · Distribution networks · Drinking water · Fate predictive modeling · Pipe relining · Quantitative chemical risk assessment (QCRA)

B. Cantoni (✉)
Department of Civil and Environmental Engineering (DICA), Environmental Section, Politecnico di Milano, Piazza Leonardo da Vinci 32, 20133 Milano, Italy
e-mail: beatrice.cantoni@polimi.it

© The Author(s) 2022   1
M. Antonelli and G. Della Vecchia (eds.), *Civil and Environmental Engineering for the Sustainable Development Goals*, PoliMI SpringerBriefs,
https://doi.org/10.1007/978-3-030-99593-5_1

1 Introduction

In recent years, the presence of micropollutants in the aquatic environment has become an issue of growing global concern. Great attention is paid to the so-called Contaminants of Emerging Concern (CECs), that belong to several families of chemicals discharged from households (e.g. pharmaceutical active compounds, estrogens), agriculture (e.g. pesticides) and industrial processes (e.g. perfluorinated compounds, alkylphenols) [1]. CECs are currently not included in routine monitoring programs, although they are potentially hazardous, as some are persistent and biologically active, even being present at extremely low concentrations in the aquatic environments (ranging from ng L^{-1} to μg L^{-1}), that made them hard to detect and quantify [2]. Anthropic activities result in direct and indirect discharge of thousands of CECs in surface water and groundwater, used as drinking water (DW) sources. Few CECs have been introduced in the revision of the European Directive on DW, while a relevant number of compounds is still unregulated or just candidate for future regulations. Moreover, the revision of the European DW Directive promoted a shift in the current paradigm, pushing the preventive estimation of human health risk, in order to identify the main risk sources and prevent and minimize risks for the consumer throughout the whole supply system (from source to tap). Human health risk prediction consists of understanding whether CECs exposure concentrations exceed a tolerable health-based threshold, derived from toxicological studies. However, the use of a risk-based approach is not easy to be achieved for CECs due to several knowledge gaps. In particular, it is hard to evaluate CECs exposure levels in DW, firstly because of their low concentrations compared to the LOQ (Limit of Quantification) values of the analytical methods, which are in continuous refining; this results in monitoring databases characterized by high percentages of censored data, i.e. data below the LOQ. Moreover, uncertain estimation of CECs exposure levels in DW is also due to a lack of consolidated engineering knowledge about their fate throughout treatment processes in drinking water treatment plants and in drinking water distribution networks. Finally, high uncertainty is related also to CECs toxicity that hinders the prioritization of CECs to be included in the regulations and also the limit to be set.

The present work aims at filling current knowledge gaps in the field of risk assessment related to CECs in DW supply systems, and at developing an holistic risk assessment approach providing an effective tool to support: (i) decision-makers in evaluating the current human health risk and prioritizing CECs regulation, (ii) water utilities in planning affordable and effective upgrading, management and/or monitoring interventions for risk minimization throughout the whole water supply system.

Actually, the present research work is crucial to achieve provision of safe DW, that is one of the major requisites for human health, related to four of the 17 Sustainable Development Goals (SDGs) of the United Nation 2030 Agenda. In particular, **SDG 6** is completely devoted to "*Ensure availability and sustainable management of water and sanitation for all*", with targets related not only to accessibility and quantity of DW, but also to its quality, as reported in target 6.3 that aims by 2030, to improve water quality by reducing pollution, eliminating dumping and minimizing release of hazardous chemicals and materials. Secondly, the provision of safe DW has high impacts on the achievement of **SDG 3**, in order to *"Ensure healthy lives and promote well-being for all at all ages"*. This is particularly clear when looking at target 3.9, aiming by 2030 at substantially reducing the number of deaths and illnesses from hazardous chemicals and air, water and soil pollution and contamination, and target 3.d, pointing at strengthening the capacity of all countries, in particular developing countries, for early warning, risk reduction and management of national and global health risks. Being the focus of this research the reduction of CECs in DW in highly-anthropized environments, this would help in *"Making cities and human settlements inclusive, safe, resilient and sustainable"*, that is **SDG 11**. Finally, the proper management of the DW production and supply system and the consequent increase in citizens trust in tap water are two key elements to achieve **SDG 12**, *"Ensuring sustainable consumption and production pattern"*. This is particularly linked to target 12.2 that promotes the achievement of sustainable management and efficient use of natural resources, and target 12.8 ensuring that people everywhere have the relevant information and awareness for sustainable development and lifestyles in harmony with nature.

2 Rationale and Methods

2.1 Problem Identification and System Conceptualization

To achieve the project goal of developing a holistic risk assessment approach to minimize health risk related to CECs in DW, the first step is the description of the system and the identification of the hazards that may influence tap water quality deterioration. The DW supply system was conceptualized in 3 elements before water is delivered to the point of use (i.e. the consumers): (i) DW sources (e.g. groundwater, surface water), (ii) drinking water treatment plants (DWTPs), and (iii) drinking water distribution network (DWDN). Thus, tap water deterioration can result from three main causes:

(1) Source contamination: the presence of CECs in source water can be due to direct or indirect discharges from different anthropic activities.
(2) Inefficiency of DWTPs towards CECs removal: currently the best available technology for CECs reduction, already present in DWTPs, consists in Granular Activated Carbon (GAC) filters. Failure of DWTPs towards CECs can be due to several factors, as low GAC-contaminant affinity ensuing low adsorption capacity, non-optimal configuration of the GAC filters in terms of dimensions and operation (in series or in parallel), and bad management with late exhausted GAC regeneration.
(3) Recontamination of treated DW through the DWDN: this could occur in case of contaminants release from pipes materials (metals, plastics) in contact with

water. This phenomenon depends on the nature of pipes materials, pipes maintenance and renovation management (e.g. substitution or relining) to prevent materials deterioration and on the effect of water aggressivity and corrosion potential on pipes.

As for the effects generated by the potential tap water contamination, a direct consequence may be the health risk for citizens, depending on the exposure levels and hazard characteristics of the analyzed CECs. On the other side, due also to the lack of knowledge and awareness of citizens about tap and bottled water quality and impacts, tap water deterioration brings citizen to drink less tap water and increase bottled water consumption. This choice has multiple negative effects, such as the exposure to health risks resulting by pollutants released by bottled water, environmental impacts due to the whole life cycle of bottled water and higher costs for citizens for drinking water.

2.2   Design of the Research

Based on the identified potential problems, experimental activity, at lab- and full-scale, advanced statistical methods and modelling techniques were combined to apportion the contribution of each element of the DW supply system in determining human health risk, in order to prioritize mitigation actions in view of an overall risk minimization. The work was structured in five tasks: (i) application of advanced method for censored CECs data analysis, (ii) development of a Quantitative Chemical Risk Assessment (QCRA) for CECs in DW supply, (iii) and (iv) predictive modelling of GAC adsorption performance towards PFAS and pharmaceuticals, respectively, and (v) modelling of bisphenol A (BPA) release from epoxy resins used for pipe relining and fate in DWDN. These tasks are interconnected according to the schematic overview given in Fig. 1 and briefly described in the following paragraphs.

3   Methods and Relevant Results

3.1   Application of Advanced Method for Censored CECs Data Analysis

Dealing with CECs, due to the low concentrations and the continuously refining of the analytical methods, monitoring databases are rich in censored data, meaning samples with concentrations below the LOQ of the analytical methods and that therefore cannot be quantified but only reported in the database as "<LOQ". Censored data are traditionally eliminated or replaced with a value between 0 and LOQ, leading to erroneous estimations, but also to not fully exploit the information contained in the monitoring database. To face this constraint, the advanced Maximum Likelihood Estimation method for left-censored data (MLE_{LC}), that combines the values above

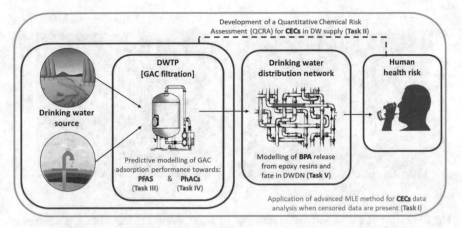

Fig. 1 Schematic overview of the project tasks and their focus on each part of the drinking water supply system

the LOQ with the information contained in the proportion of censored data, was here proposed and tested [3]. A field monitoring campaign was designed to evaluate CECs concentrations in addition to routinely monitored parameters in groundwater (GW) and DW in a highly urbanized area. A database was built with data of 19 contaminants (metals, volatile organic compounds, pesticides and perfluorinated compounds) in 5,362 GW and 12,344 DW samples, collected from 2012 to 2017 in 28 DWTPs of the monitored urbanized area. The MLE_{LC} was applied to estimate the statistical distribution of CECs concentrations and results were compared to the traditional methods. Three applications for the comparison were selected, that are fundamental to predict the future raw water quality and to define intervention scenarios: evaluation of contaminants concentration time trend, estimation of treatment removal efficiency and risk assessment. Finally, a guideline was provided to select the data elaboration method to be preferred based on the comparison of the methods estimation errors, as a function of the percentage of censored data (from 0.3 to 99.0%) and the amplitude of concentration data range. This was made possible by the wide range of contaminants, the several DWTPs and the numerous sampling locations considered in this study.

An example of this task output is reported in Fig. 2, where the estimation error of the health risk likelihood (e_L) for the elimination and MLE_{LC} methods is reported as a function of both the percentage of the censored data and the amplitude of data range, reported as the ratio between the LOQ and the concentration data 95th percentile.

The MLE_{LC} method was demonstrated to be the most accurate method with estimation errors always below 20%. On the other hand, traditional elimination and substitution methods can lead to erroneous conclusions under- or overestimating the human health risk, especially for high percentages of censored data. This is of particular interest for CECs, often characterized by high censored percentages but with severe effects on human health also at very low concentrations. An accurate estimation of their risk is necessary to correctly plan the upgrading interventions of current

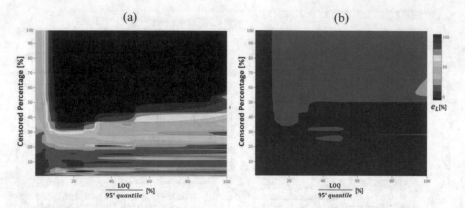

Fig. 2 Contour plot of the estimation error (e_L) of health risk as a function of the percentage of censored data and amplitude of data range for elimination **a** and MLE$_{LC}$ **b** methods [3]

DWTPs in order to meet the new regulatory limits proposed worldwide for CECs. In fact, both an underestimation or an overestimation of the exceedance probability could have important drawbacks in terms of underestimation of the risk or overestimation of the intervention needed, and the related costs, leading to potentially not precautionary or unsustainable intervention plans.

3.2 Development of a Quantitative Chemical Risk Assessment (QCRA) for CECs in DW Supply System

Currently, chemical risk assessment (CRA) in DW applications is deterministically performed and uncertainties are taken into account by selecting conservative point values, such as a high exposure concentration, or a lower bound estimate of the health-based guideline level [4]. Then, the ratio between the exposure concentration and the health-based guideline level point values is calculated as the deterministic benchmark quotient (BQ) that provides an indication of no risk or risk for BQ below or above 1, respectively [5]. However, evaluating the risk for CECs in DW is not easy due to several knowledge gaps and high uncertainties related both to their exposure levels and hazard characteristics. Therefore, in this study [6], a new probabilistic procedure, that is the quantitative chemical risk assessment (QCRA), was developed including in risk calculation the uncertainties in both exposure and hazard assessments, obtaining as output the BQ probabilistic distribution, useful to estimate the probability of BQ of exceeding threshold value of 1, P(BQ > 1).

As for the exposure assessment, the probabilistic distribution of CECs concentration in DW was estimated based on their concentration in source water and simulating the breakthrough curves of GAC adsorption filters, through the Ideal Adsorbed Solution Theory (IAST) model. IAST model was implemented by AquaPriori, a Python based treatment simulation tool developed by KWR (Utrecht, NL), that was upgraded to accept distributions—instead of point values—for input parameters. The uncertainties in the CECs hazard assessment were included using the APROBA-Plus tool developed by the USEPA, as described by [4].

The model inputs and output uncertainties were evaluated by sensitivity and uncertainty analyses for each step of the risk assessment to identify the most relevant factors affecting risk estimation, highlighting future research needs to improve reliability of risk assessment. To stress the potential of this new QCRA approach, several case studies and GAC management options were considered, with a focus on BPA as an example CEC. The probabilistic risk quantified by the QCRA was compared to the deterministic one estimated by the traditional CRA. An example of how using the developed QCRA procedure to manage and optimize CECs treatment in the DWTP is provided in Fig. 3, where the risk estimation output of the deterministic CRA and QCRA are displayed as a function of two operating conditions of the GAC filters: (i) the Empty Bed Contact Time (EBCT), that is the time the DW is in contact with the GAC in the filter, and (ii) GAC regeneration time (BV_{REG}), that is the time when the exhausted GAC is treated to desorb contaminants and restore its adsorption capacity.

The EBCT has a negligible influence on the human health risk, compared to the regeneration time. Therefore, the EBCT seems not to be a relevant parameter through which optimize BPA removal by GAC process, suggesting that engineered intervention to increase EBCT does not imply any significant risk reduction. Both the deterministic BQ and the probability of exceeding BQ equal to 1 ($P(BQ > 1)$) decrease with the reduction of the regeneration time. However, using the deterministic approach (Fig. 3a), a regeneration time equal to 25,000 BV (bed volumes) would be selected as the optimal value under which there is no significant risk ($BQ_{DET} = 1$), while including all the uncertainties within the QCRA it comes out that there is still a 12.0% probability of exceeding BQ value equal to 1 (Fig. 3b) and the optimal

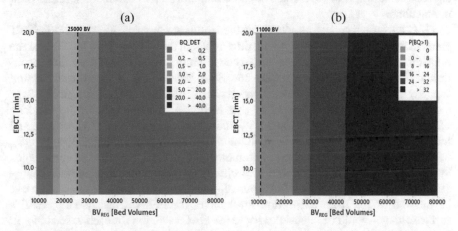

Fig. 3 Contour plot of BQ_{DET} **a** and $P(BQ > 1)$ **b** as a function of the empty bed contact time (EBCT) and the regeneration time (BV_{REG}) [6]

regeneration time of 11,000 BV would result in a virtual 0% P(BQ) > 1. Therefore, the deterministic BQ provides the less precautionary approach, which does not identify residual human health risk in specific cases. Actually, based on the deterministic BQ, costs associated to the regeneration time adopted as optimal threshold would lead to benefits lower than expected. In conclusion, the QCRA is more effective than deterministic CRA in evaluating the effect of each management option in risk minimization, permitting to select and prioritize the most appropriate interventions.

3.3 Modeling of GAC Adsorption Performance Towards CECs

Since with the QCRA it was found that modelling of GAC breakthrough curves has a relevant role in the accuracy of risk estimation [6], a thorough experimental work has been designed and performed to more accurately model GAC performance towards CECs, in particular pharmaceutical active compounds (PhACs) and perfluoroalkyl substances (PFAS) (for a detailed description see [7]). Four commercial GACs were tested by both isotherm batch experiments and rapid small-scale column tests (RSSCT), to calibrate CECs breakthrough curves. Experiments were performed on 8 PFAS and 8 PhACs in 3 water matrices, which were tap water and additional two synthetic matrices at lower dissolved organic carbon (DOC) and two levels of conductivity. As for activated carbon, 4 GACs were tested, differing for origin (bituminous or coconut-based), surface charge (neutral or positively charged), number of reactivation cycles (virgin and reactivated GACs) and porosity (micro- and mesoporous). Results were explored through multivariate analyses (i.e. factorial and cluster analyses) and used to calibrate a performance model able to predict the breakthrough curves as a function of CECs, activated carbon and water characteristics, and their interactions.

GAC performance can be related to the isotherm constant, K_F, and the RSSCT parameter BV_{50}, corresponding to the time at which the CEC reaches the 50% of the breakthrough curve, having a filter outlet concentration equal to half the inlet concentration (Cout/Cin = 0.5). Higher K_F and BV_{50} values indicate greater adsorption capacity of the adsorbent. In general, a good agreement among isotherm and RSSCT results was outlined, as suggested by data in Fig. 4, showing that for the majority of the analyzed PFAS, both K_F and BV_{50} increase from short-chain hydrophilic and marginally hydrophobic PFAS to medium-chain and hydrophobic PFAS, to PFOS.

In addition to the confirmation of literature evidences on the effect of compounds hydrophobicity on GAC adsorption capacity, it was found that GAC surface charge affects performance more than GAC porosity, therefore electrostatic interaction can be inferred as the main adsorption mechanism for PFAS. Consequently, GAC's surface charge should be checked prior to the GAC selection for each case-study. In fact, considering the pH of the water to be treated, positively-charged GAC should

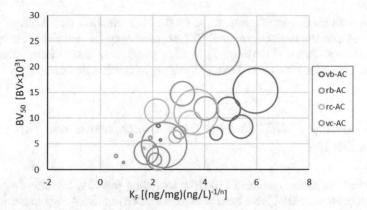

Fig. 4 Bubble chart for PFAS removal by adsorption on activated carbon (AC). Bubble centers are located according to the isotherm K_F and the RSSCT BV_{50}; bubble diameters are proportional to respective PFAS hydrophobicity [7]

be preferred rather than negatively-charged or neutral GAC, to fully exploit electrostatic attraction towards negatively-charged PFAS [8]. Moreover, the interaction between CEC hydrophobicity and GAC porosity was found to significantly affect performance. Therefore, GAC selection should also consider PFAS mixture in the source water to be treated.

Finally, a correlation was built between the reduction of UV absorbance at 254 nm (UVA_{254}), that is an easily measurable parameter also by on-line sensors, and CECs removal (Fig. 5), in order to evaluate whether UVA_{254} can be used as a proxy variable for CECs continuous on-line monitoring.

Actually, UVA_{254} removal is correlated to CECs removal, independently from the type of GAC and water matrix. Therefore, CECs-UVA_{254} correlations can be easily

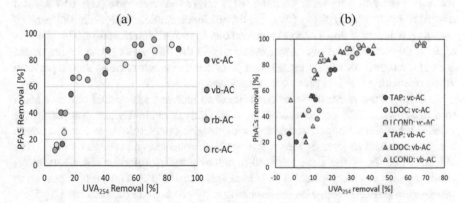

Fig. 5 Correlations between: **a** overall PFAS removal and UVA_{254} removal as a function of the four tested ACs [7], **b** overall PhACs removal and UVA_{254} removal as a function of two tested ACs and three water matrices

obtained with batch experiments, to be used in combination with UVA_{254} on-line monitoring data to predict the overall CECs breakthrough in real full-scale systems. This is an important tool to promptly identify possible system failures, which may result in human health risk, and to rapidly apply mitigation measures.

3.4 Modeling of BPA Release from Epoxy Resins and Fate in DWDN

Monitoring and management of DWDNs, including possible leaching from materials in contact with DW, have been stressed as crucial to avoid re-contamination of drinking water leading to a potential increase of human health risk. Recent scientific studies and regulations clearly highlighted the leaching of BPA from plastic materials used to renovate DWDNs pipelines as one of the major hazardous events, resulting in severe consequences for human health. Therefore, to complete the "from source to tap" holistic risk assessment, potential recontamination events in the DWDN should be properly studied. Here, a sound approach to evaluate BPA release from epoxy resins used to renovate pipelines, comprising experimental and modeling work, is reported [9].

Lab migration tests were performed on three commercial epoxy resins, characterized by different formulations, BPA content and installation techniques, put in contact with three water matrices: (i) tap water, (ii) deionized water, (iii) chlorinated tap water. The most critical resin was selected for a second set of experiments designed with the Design of Experiments (DoE) method, in order to build a BPA release model as a function of two water characteristics that were varied in a realistic range: (i) chlorine concentration, from 0 to 0.4 $mgCl_2/L$, (ii) water stability, described by the aggressivity index (AI) varying from 11.5 to 13.5. Tests lasted about 170 days to account for both short and long-term leaching. BPA migration over time (Fig. 6) was well described by a combination of two 1st-order kinetic models with an initial peak of leaching, a decrease and a second increase due to resins' deterioration. The analysis of BPA migration trend over time, especially in the first week provides important insights on monitoring and management practices that water utilities should perform when renovating DWDN pipes with epoxy resins.

Looking at the effect of water conditions, an increase of residual chlorine leads to a decrease in BPA concentration in water. This is evident when comparing the trend in tap water without chlorine and with chlorine concentration at 1 mg/L, where the initial peak is absent. However, it is not BPA release to be reduce by increasing chlorine, but likely the BPA released is actually transformed in chlorination by-products, such as chloro-phenols and chloro-bisphenols [10]. Thus, it is important to monitor not only BPA but also its chlorination by-products in order to have a proper evaluation of the overall human health risk due to the mixture of chemicals. As for the water stability, aggressive waters display higher BPA release, but the effect is

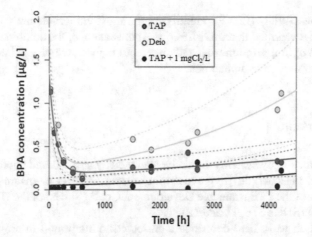

Fig. 6 BPA migration over time: dots correspond to experimental data as a function of the tested water matrix. Solid lines indicate the estimated models, dashed lines represent the models 95% confidence intervals [9]

more evident when looking at the total mass released, compared to the initial peak, for which the chlorine concentration has greater effect.

To evaluate the fate of BPA in the DWDN, the validated BPA release model was combined with the hydraulic model of a portion of a DWDN in a highly urbanized area, where two epoxy resins are installed. EPANET MSX software was used. A field monitoring campaign was also designed to measure BPA concentrations in locations nearby pipelines renovated with epoxy resins, for model validation. The model was successfully validated on full-scale monitoring data, demonstrating the occurrence of BPA leaching and the potential risk for consumers, especially if appropriate re-opening procedures are not adopted. The model allowed to simulate the current fate

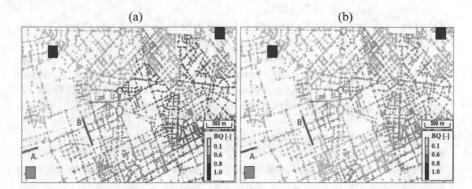

Fig. 7 Estimated BQ due to released BPA concentration in a portion of a DWDN where two pipes were renovated by epoxy resins relining (blue segments) as a function of the time from installation: **a** after 1 **day**; **b** after 30 days [9]

of BPA in the DWDN (Fig. 7), identifying the most vulnerable areas, with higher
potential risk, described in terms of BQ; as a consequence, the combined model can
be adopted to optimize monitoring and intervention plans, which can be site-specific
customized to minimize human risk.

4   Conclusions

Drinking water supply systems should be managed by a risk-based approach to evaluate potential risks and devise preventive measures, in order to ensure the achievements of four of the 17 Sustainable Development Goals (SDGs) of the United Nation
2030 Agenda (SDGs 3, 6, 11 and 12).

The research work here described contributed to adapt and improve the application of risk assessment to control the spread of CECs in DW supply systems,
overcoming some of the existing knowledge gaps and handling high uncertainties
which characterize both CECs presence and toxicity and their removal. The work
performed, integrating experimental and modeling features, has permitted to provide
a supporting tool for:

– Water utilities, that need to verify whether their current control measures are
  adequate to meet new regulatory limits for CECs in DW and to elaborate intervention plans to effectively minimize human health risk. In this context, the QCRA
  can be used to apportion the contribution of each step of the supply system
  to overall risk minimization, to prioritize the interventions. The application of
  QCRA in combination with the simulation of GAC performance under different
  operating conditions allows optimizing GAC filters up-grade and management.
  Finally, several practical indications were provided both on the criteria for GAC
  and epoxy resins selection, and on planning monitoring campaigns to evaluate
  CECs removal by GAC filters and to minimize recontamination events for DW
  contact with materials in DWDNs.
– Decision makers, that need to prioritize CECs to be included in regulations, beside
  the high uncertainty involved in this process. In this context, the QCRA can be
  effectively used to assess the exposure and hazard of different CECs, including
  the related uncertainties, to evaluate which CECs pose the highest risk for human
  health. Moreover, the study on epoxy resins could provide important tools for
  decision makers to accurately regulate the characteristics of materials in contact
  with DW.
– Scientific community, that needs to fill the knowledge gaps and reduce the uncertainties related to CECs exposure in DW and resulting health effect. In this
  context, this work provided indications on how to reduce estimation errors in
  different applications when analyzing databases characterized by high percentages
  of censored data. Additionally, the sensitivity and uncertainty analyses applied in
  combination to the QCRA can be useful to identify future needed research investigations and directions to reduce uncertainties in risk estimation. Finally, this

work showed how to combine experimental tests at different scales, prediction modelling tools and field monitoring to reduce uncertainties related to CECs fate in GAC system and DWDN.

Acknowledgements I would like to acknowledge my supervisor Prof. Manuela Antonelli for her support throughout the whole Ph.D. and all the researchers at Politecnico di Milano, German Umweltbundesamt, KWR research center, and RIVM for their fruitful collaboration. This research was funded by Metropolitana Milanese S.p.A. (MM). I would like to thank Fabio Marelli and MM laboratory staff for their helpful cooperation.

References

1. Stuart, M., Lapworth, D., Crane, E., Hart, A.: Review of risk from potential emerging contaminants in UK groundwater. Sci. Total Environ. **416**, 1–21 (2012)
2. Valcárcel, Y., Alonso, S.G., Rodríguez-Gil, J.L., Maroto, R.R., Gil, A., Catalá, M.: Analysis of the presence of cardiovascular and analgesic/anti-inflammatory/antipyretic pharmaceuticals in river- and drinking-water of the Madrid Region in Spain. Chemosphere **82**, 1062–1071 (2011). https://doi.org/10.1016/j.chemosphere.2010.10.041
3. Cantoni, B., Delli Compagni, R., Turolla, A., Epifani, I., Antonelli, M.: A statistical assessment of micropollutants occurrence, time trend, fate and human health risk using left-censored water quality data. Chemosphere **257**, 127095 (2020)
4. Bokkers, B.G.H., Mengelers, M.J., Bakker, M.I., Chiu, W.A., Slob, W.: APROBA-plus: a probabilistic tool to evaluate and express uncertainty in hazard characterization and exposure assessment of substances. Food Chem. Toxicol. **110**(October), 408–417 (2017)
5. Baken, K.A., Sjerps, R.M.A., Schriks, M., Van Wezel, A.P.: Toxicological risk assessment and prioritization of drinking water relevant contaminants of emerging concern. Environ. Int. **118**(May), 293–303 (2018)
6. Cantoni, B., Penserini, L., Vries, D., Dingemans, M., Bokkers, B., Turolla, A., Smeets, P., Antonelli, M.: Development of a quantitative chemical risk assessment (QCRA) procedure for contaminants of emerging concern in drinking water supply. Water Res. **194**, 116911 (2021)
7. Cantoni, B., Turolla, A., Wellmitz, J., Ruhl, A.S., Antonelli, M.: Perfluoroalkyl substances (PFAS) adsorption in drinking water by granular activated carbon: Influence of activated carbon and PFAS characteristics. Sci. Total Environ. 148821 (2021)
8. Xiao, F., Zhang, X., Penn, L., Gulliver, J., Simcik, M.: Effects of monovalent cations on the competitive adsorption of perfluoroalkyl acids by kaolinite: Experimental studies and modeling. Environ. Sci. Technol. **45**(23), 10028–10035 (2011)
9. Cantoni, B., Cappello Riguzzi, A., Turolla, A., Antonelli, M.: Bisphenol A leaching from epoxy resins in the drinking water distribution networks as human health risk determinant. Sci. Total Environ. **783**, 146908 (2021)
10. Bruchet, A., Elyasmino, N., Decottignies, V., Noyon, N.: Leaching of bisphenol A and F from new and old epoxy coatings: laboratory and field studies. Water Sci. Technol. Water Supply **14**(3), 383–389 (2014)

Influence of the Collection Equipment on Organic Waste Treatment: Technical and Environmental Analyses

Giovanni Dolci

Abstract The research investigated the influence of different collection bag types on the environmental and energy performances of the food waste management chain, comparing paper and bioplastic bags. First, the use of bags during the food waste household storage was examined. Subsequently, the behavior of bags when subjected to anaerobic digestion was evaluated, performing Biochemical Methane Potential tests and semi-continuous co-digestion tests with the food waste, to simulate the operating conditions of full-scale digesters. Finally, the performances of the food waste management chain were evaluated, with a Life Cycle Assessment (LCA). The experimental tests showed a more favorable behavior of paper bags, showing a very good compatibility with the anaerobic digestion. The LCA results revealed how paper bags lead to improvements in the impact associated to the food waste management.

Graphical Abstract

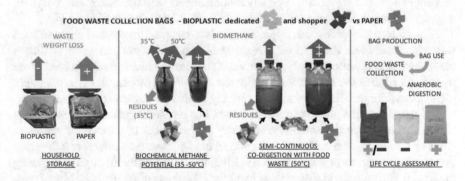

Keywords Food waste · Anaerobic digestion · Life cycle assessment · Affordable and clean energy · Sustainable cities and communities · Responsible consumption and production

G. Dolci (✉)
Department of Civil and Environmental Engineering, Politecnico di Milano, Piazza Leonardo da Vinci 32, 20133 Milan, Italy
e-mail: giovanni.dolci@polimi.it

1 Introduction

The organic fraction is generally the most relevant among all the separately collected materials in the municipal solid waste. In Italy, 6.4 million metric tons of organic waste were separately collected in the year 2019 [1]. In addition to the typical components of this fraction (food waste and green waste), the amount of compostable bioplastics conferred within the organic waste has grown to 3.9% of the organic waste in the period 2019/2020 compared to 1.5% in the period 2016/2017 [2].

Bioplastics show several issues when managed together with the food waste. Such criticalities are first associated with the mechanical pre-treatments that precede the biological process: when the waste is subjected to size-based separation, most of the bioplastic products are discarded as residues, similarly to conventional plastics [2, 3]. This is particularly relevant in anaerobic digestion (AD) plants, where pre-treatments are often very intense [3] to avoid further hydraulic and operational problems. The problems and costs associated with the management of discarded bioplastic items are amplified by the fact that, when removed, they drag a non-negligible amount of food waste that remains adhered to them and is not delivered to the AD process [2, 3].

Although the increasing amount of different bioplastic items, a relevant contribution is still constituted by the bags employed for the collection of food waste. In Italy, the current collection systems of food waste from households are mainly based on the use of bioplastic bags. In detail, the collection can be performed with bags specifically sold for this purpose (dedicated bags) or with bags used for the overall shop at the supermarkets (shopper bags) that can be reused for the food waste separate collection. Both types are typically manufactured with the Mater-Bi® polymer, a compostable bioplastic according to the UNI EN 13,432:2002 standard [4].

Alongside, a less widespread type of paper bag designed for the collection of food waste is available on the market. It is made of recycled fibers and composed by a main bag and a separate cartonboard bottom to be inserted inside the main bag before its use. As regards the compatibility with the AD, unlike bioplastic bags, the paper does not require prior removal since it quickly breaks down during pre-treatments.

Building up on the previous considerations, this research was carried out with the aim to analyze the environmental and energy performances of the overall treatment chain of the food waste collected from households, focusing on how it is influenced by the different types of collection bags. First, the use of different bag types during the food waste household storage was examined. Subsequently, the behavior of bags when subjected to AD was evaluated. Finally, the influence of the bag types on the environmental performances of the food waste management chain was evaluated, by means of the Life Cycle Assessment (LCA) methodology.

First, the aim at evaluating the collection bag type that optimizes the household management and the subsequent collection of food waste agrees with the **SDG 11**, promoting the development of more sustainable cities and communities. In parallel, the goal to reduce the potential impacts of the overall food waste management chain, promoting the production and the use of more sustainable collection bags, addresses the **SDG 12**, supporting more sustainable productions and consumptions. Finally, the aim at identifying the management option allowing to increase the energy valorization of food waste through the AD process agrees with the **SDG 7**, promoting the production of energy both sustainable and widely available wherever food waste is produced.

2 Materials and Methods

2.1 Household Storage Tests

Firstly, the behavior of the food waste during the household storage when collected inside the different types of bag was examined. In detail, the food waste weight loss during the household storage (i.e. time occurring between the delivery in the bag by the user and its collection) was analyzed by adopting a dynamic, progressive bag filling. This approach aimed at investigating the progressive bag filling due to the daily food consumption, differently from the typical methodology applied in tests reported in the literature, where the bag is completely filled at the beginning.

In two years, 112 domestic tests were performed in parallel to compare paper and bioplastic bags behavior: 59 paper vs. bioplastic dedicated bags and 53 paper vs. bioplastic shopper bags. In each comparative test, one paper bag and one bioplastic bag were placed inside aerated bins. Before each bag filling (twice a day, after lunch and dinner) the food waste was homogenized and split into two portions with the same weight discharged respectively in the paper bag and in the bioplastic bag. Each test lasted 5 days. At the end of the test, the two bags were removed and weighed. The weight loss with respect to the total inserted waste was then calculated for both bags. Subsequently, the differences in terms of weight loss between the two materials were statistically tested (Mann-Whitney U test). In addition, empirical observations on the resistance of the bags were performed during the tests.

Six bags (three dedicated and three shopper bags) manufactured by different companies were tested. The analyses were performed during the different seasons, with the aim to consider the variations of both the environmental conditions (temperature and humidity) and the composition and characteristics of the food waste. The tests were conducted by different households in order to consider various eating habits and therefore different amounts and characteristics of the generated food waste.

2.2 Evaluation of the Anaerobic Degradation of Food Waste Collection Bags

In the second part of the study, the treatment stage of the food waste management chain was examined. Assuming that the operational problems associated to the management of bioplastics in biological plants (see Sect. 1) could be solved in the future, bioplastic collection bags must be compatible with biological processes. According to the UNI EN 13,432: 2002 technical standard, only aerobic degradability tests must be performed, while it is generally not necessary to test the anaerobic biodegradability [4]. However, in Italy there is an increasing tendency to manage food waste in AD plants; therefore, it is essential to verify the behavior of bags even under such conditions.

Accordingly, preliminary biochemical methane potential (BMP) tests were performed to evaluate the anaerobic degradability and the corresponding biomethane yield of the three bag types (paper bag—PB, one bioplastic dedicated bag—BDB, and one bioplastic shopper bag—BSB). For the tests, bags were manually cut in square pieces of 1 cm side. Tests, performed both under mesophilic (35 ± 0.5 °C) and thermophilic conditions (50 ± 0.5 °C), were carried out with digestates, serving as inoculum, sampled from full-scale AD plants processing food waste. An inoculum to substrate ratio equal to 2 $VS_{inoculum}/VS_{substrate}$ was adopted. A mineral medium with macro and micro-nutrients was also dosed before the tests.

According to the results of preliminary BMP tests (see Sect. 3.2), a deeper investigation on bags anaerobic degradability under thermophilic conditions was performed with new lab-scale tests. Four bioplastic bags were selected, including two dedicated bags (bioplastic dedicated bag 1—BDB1 and bioplastic dedicated bag 2—BDB2), and two shopper bags (bioplastic shopper bag 1—BSB1 and bioplastic shopper bag 2—BSB2). Bags manufactured by different companies with different thickness (two shoppers and two dedicated bags) and colors (red, yellow, not colored, and green) were selected. Moreover, the type of paper bag (PB) examined in household storage tests (see Sect. 2.1) and in preliminary BMP assays was tested.

The experimental plan first included BMP tests, performed on all five bags, and on a synthetic food waste (its composition was defined based on 90 composition analysis of the organic fraction received in composting plants) at 50 ± 0.5 °C.

Subsequently, to better simulate the real operating conditions of full-scale digesters, collection bags were subjected to semi-continuous co-digestion tests with the food waste. To the authors' knowledge, similar tests are reported only in one literature study, though performed only under mesophilic conditions.

Tests were performed on two out of the four bioplastic bags and on the paper bag (co-digestions of food waste—BSB1, food waste—BDB1, and food waste—PB), in 2.4 L stirred reactors, under thermophilic conditions (50 ± 0.5 °C). Semi-continuous conditions were obtained by removing part of the digestate and by adding the new substrates and water twice a week, to maintain a hydraulic retention time (HRT) of 21 days and an organic loading rate (OLR) of 2.20 kg COD/($m^3 \times$ d) (COD, Chemical Oxygen Demand). Mineral mediums with macro and micro-nutrients were prepared and periodically dosed in the reactors.

Tests were performed in four reactors; in the first period (phase 1, 20 feed cycles corresponding to more than three HRTs) all the bottles were fed with only food waste to reach inoculum acclimation and stationary conditions. In the second period (phase 2, 19 feed cycles), three reactors were also fed with bioplastics (11.5% of the OLR on COD basis), the fourth serving as blank. The selected proportion corresponds to about 1 kg of food waste inserted into a collection bag.

Statistical tests (Kruskal-Wallis and Mann-Whitney U tests) were applied for the evaluation of differences among reactors in terms of methane production. First, reactors were compared in the last part of phase 1 to verify the absence of statistically significant differences before the phase 2. Moreover, the differences in phase 2 were evaluated.

In phase 2, the extracted digestates were sieved (2 mm) to recover undigested pieces of bags. All the residual bioplastic pieces were washed with water, dried at 35 °C, and weighed to evaluate their mass losses during the digestion. Moreover, undigested pieces with a surface equal to at least ¾ of that of the input were recovered and counted.

2.3 Life Cycle Assessment

The environmental performances of the management chain of the food waste collected from households were evaluated, by means of the LCA methodology, comparing two systems in which the employed collection bags are respectively made of bioplastic and paper. Differently from the LCAs reported in the literature for the evaluation of the potential impacts of the food waste management chain, this analysis included stages such as the collection bag production (and its influence on the management system) and the food waste household storage. *The management of 1 kg of food waste generated (i.e., inserted into the collection bag) at the household* was assumed as functional unit. Non-compostable materials mistakenly discarded along with food waste were excluded because their quantity was assumed not to be affected by the different bags. The system boundary included the overall food waste management chain (Fig. 1).

Potential impacts were calculated by examining 16 midpoint impact categories with the indicators and assessment models of the Environmental Footprint 2.0 method [5]. Normalization and weighing were applied with the factors considered in this method.

The investigated systems were mainly modeled with primary inventory data for the Italian context as regards the paper bag manufacturing, the bags distribution, the bags use (modeled according to results of the household storage tests, see Sect. 3.1), and the food waste treatment by means of the AD process.

The ecoinvent database (version 3.5 *allocation, cut-off by classification* system model) was used to support the modeling [6]. The SimaPro software (9.0 version) supported the analysis. For the modeling of the life cycle stages of products included in the systems (collection bags and packaging), two different approaches were considered: the approach applied in the International EPD system [7] and the one applied in the Product Environmental Footprint (PEF) methodology [8, 9].

Firstly, a *Baseline scenario* was modeled, representing the average situation of the current Italian food waste management system, strongly based on bioplastic bags, in terms of bag filling level and frequency of food waste collection. These conditions were assumed to be identical with the alternative use of paper bags (Table 1).

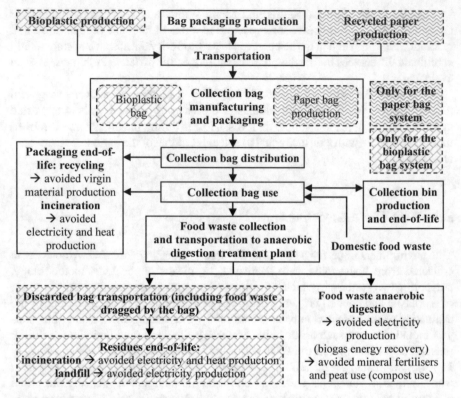

Fig. 1 Stages evaluated in the life cycle assessment of the two food waste management systems

A *Sensitivity scenario* was then examined (Table 1), considering the worst conditions for the system based on bioplastic bags, to evaluate the maximum benefits achievable with the use of paper bags. In particular, the bioplastic bag system shows the worst performances when:

• bioplastic bags are filled with only 1 kg of food waste (e.g., waste generated by one single person) and therefore collected, not completely filled, twice a week;
• bioplastic bags discarded during pre-treatments of the AD drag an amount of food waste higher than the average plant performances, and residues are sent to landfill.

The worst performances of the current system were assumed to be improved with the use of paper bags. As detailed in Sect. 3.1, the use of paper bags allows for a higher weight loss and a lower generation of leachate and odor during the household storage. Accordingly, when the amount of generated waste is low, paper bags could be employed for a longer time increasing the bag filling level (from 1 to 2 kg), which allows to decrease the collection frequency (from bi-weekly to weekly).

Table 1 Parameters considered in the *Baseline scenario* and in the *Sensitivity scenario* of the life cycle assessment

Parameter	Baseline scenario		Sensitivity scenario	
	Bioplastic bag system (dedicated and shopper)	Paper bag system	Bioplastic bag system (dedicated and shopper)	Paper bag system
Amount of food waste inserted into each bag (kg)	2	2	1	2
Food waste collection frequency	Bi-weekly	Bi-weekly	Bi-weekly	Weekly
Mass fraction of food waste dragged with bioplastic bags removed during pre-treatments of anaerobic digestion and subsequent treatment	2% (dedicated) / 3% (shopper) sent to a waste-to-energy plant together with bioplastic bags	–	10% sent to landfill together with bioplastic bags	–

3   Results and Discussion

3.1   Household Storage Tests

Bioplastic and paper collection bags showed a different behavior during the household storage tests, with the paper allowing for higher weight losses: +29% and + 44% on average, compared respectively to bioplastic dedicated and shopper bags (Fig. 2).

The higher weight losses of the food waste collected inside paper bags are favored by the breathable fabric of paper that allows for a relevant evaporation of moisture. According to the results, the weight losses of the food waste collected inside paper and bioplastic bags are statistically different.

In addition, paper bags allow for a lower odor and leachate release during their use at the household, paving the way to a potential decrease of the frequency of food waste collection, thus reducing costs and environmental impacts.

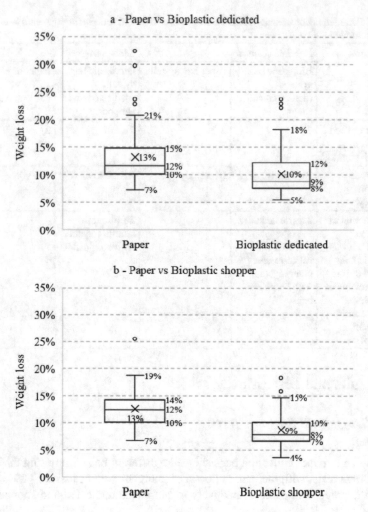

Fig. 2 Household storage tests results: food waste weight losses in the time between the delivery in the bag by the user and the collection for the 59 comparative tests paper versus bioplastic dedicated bags (**a**) and the 53 comparative tests paper vs bioplastic shopper (**b**)

3.2 Evaluation of the Anaerobic Degradation of Food Waste Collection Bags

The preliminary BMP assays showed a very limited anaerobic degradability of bioplastic bags (9–15%) under mesophilic conditions, with residues of the substrates at the end of tests showing only slight changes in color brilliance, without any appreciable size reduction compared to the input samples (Fig. 3). In the thermophilic tests, the bioplastic bags showed a degradability in the range 22–57%. On the contrary,

Fig. 3 Undigested bioplastic bags samples after the mesophilic preliminary biochemical methane potential tests

Bioplastic shopper bag (BSB)   Bioplastic dedicated bag (BDB)

Table 2 Biochemical methane potential (BMP) and semi-continuous tests result: anaerobic degradability on Chemical oxygen demand (COD) basis of tested substrates, evaluated considering a theoretical production of 330 NmLCH₄/gCOD (about 6% of COD for growth)

Substrate	Anaerobic degradability	
	BMP test (duration of test)	Semi-continuous co-digestion test
Bioplastic shopper bag 1 (BSB1)	84% (41 days)	12%
Bioplastic shopper bag 2 (BSB2)	87% (56 days)	Not tested
Bioplastic dedicated bag 1 (BDB1)	71% (41 days)	27%
Bioplastic dedicated bag 2 (BDB2)	93% (45 days)	Not tested
Paper bag (PB)	74% (23 days)	82%
Food waste	98% (21 days)	92%

paper bags showed a good anaerobic degradability under both temperature conditions (58–66%) without residues of substrate remaining after the tests.

The subsequent BMP tests under thermophilic conditions on four commercial types of bioplastic bags indicate a good degradability (>71%) of all these substrates, without residues after the tests, although obtained after a very long time (Table 2). On the contrary, the degradation of paper is much faster, with 90% of the final BMP reached in six days. Similarly, for the food waste, two and six days are enough to reach 50% and 90% of the final BMP, respectively.

As regards the semi continuous co-digestion tests, the following reductions in terms of methane production were observed in reactors fed with bags pieces, compared to the reactor fed with only food waste:[1] −9.9% (BSB1 + food waste), − 8.0% (BDB1 + food waste), and −1.2% (PB + food waste). According to statistical tests, the differences in methane productions resulted statistically significant between

---

[1] All the results of semi-continuous tests are related to the second part of phase 2 (starting 40 days after the first introduction of bag pieces into the reactors).

each of the two reactors fed with bioplastic bags and the reactor fed with only food waste and between each of the two reactors fed with bioplastic bags and the reactor fed with the paper bag.

The reduced methane production of bioplastics corresponds to a low anaerobic degradability (<27%; Table 2). On the contrary, very interesting perspectives are offered by the tested paper bag, since its anaerobic degradability in the semi-continuous tests (82%) resulted even higher than that observed in the BMP tests (74%; Table 2). This indicates a very good compatibility with the AD process, suggesting effects of biomass acclimation or synergic effects given by the co-digestion of PB and food waste.

As regards the physical status of undigested substrates, several undigested pieces of BSB1 and BDB1, similar in shape and color to the fed pieces, were observed (Fig. 4).

The overall mass of undigested pieces resulted equal to 93% and 69% of the weight inserted for BSB1 and BDB1, respectively. The corresponding weight losses agree with the different anaerobic degradability observed for the two bioplastics (Table 2). The similarity in shape between fed and undigested bioplastics is confirmed by the very high number of residual pieces with a surface equal to at least ¾ of that of fed substrate, corresponding to 96% and 98% for BSB1 and BDB1, respectively.

Fig. 4 Pieces fed to semi-continuous tests of bioplastic shopper bag 1—BSB1 (**a**), bioplastic dedicated bag 1—BDB1 (**b**), and paper bag—PB (**c**). Undigested pieces of semi-continuous tests of BSB1 (**d**), BDB1 (**e**), and PB (**f**)

As regards PB, only very small amounts of residues were retained during sieving, in which single pieces were not detectable (Fig. 4).

3.3   Life Cycle Assessment

The LCA results show a relevant influence of the collection bag on the potential impacts of the food waste management chain. The comparison highlighted a beneficial influence associated with the use of recycled paper bags instead of bioplastic bags, in particular the dedicated ones; shopper bags are less impacting because they are used, as the first purpose, for carrying the grocery shopping. Accordingly, only 50% of impacts related to the production and the treatment at the end of life of the bioplastic shopper bag are included in the analyzed system. The benefits of the use of paper bags are associated to both the bag manufacturing (less impacting, especially thanks to the use of recycled fibers) and the benefits in the AD treatment, since they are not discarded during pre-treatments, differently from bioplastic bags. Anyway, the methodological approach used in the LCA modeling has an important influence on the comparison. In particular, the paper bag system achieves the highest environmental advantages with the EPD approach (Fig. 5a), while using the PEF entails a significant increase of the impacts because of the effect of partially considering the virgin paper production, in place of the use of recycled fibers, with an important influence on the comparison between paper and bioplastic bags, as shown in Fig. 5b.

Table 3 shows the potential impacts of both the food waste management systems after the normalization and the weighing stages.

Examining the *Baseline scenario*, with the EPD approach, the paper bag system allows for a decrease of the potential impacts compared to the bioplastic system, both considering dedicated and shopper bags. Moreover, it is important to underline that the paper bag system is characterized by a result negative in sign (i.e., the management of food waste collected inside paper bags as modeled allows for environmental benefits).

With the PEF approach, the paper bag system still allows for a decrease of potential impacts compared to the bioplastic dedicated bag system (−57%). On the contrary, the bioplastic shopper bag system is better (−34%) than the system based on paper bags.

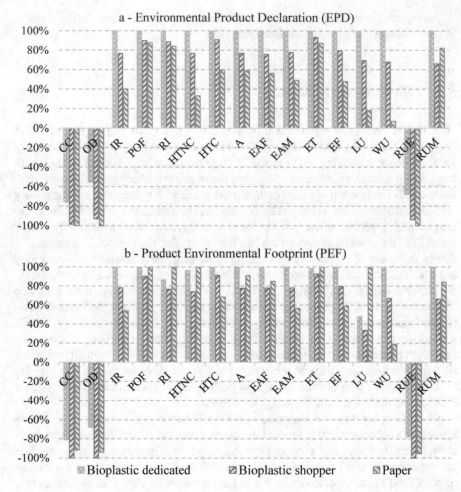

Fig. 5 Life Cycle Assessment results: comparison of potential impacts of the paper bag and the bioplastic bag systems, with the EPD (**a**) and the PEF (**b**) approaches. Impact categories: CC Climate change; OD Ozone depletion; IR Ionizing radiation, human health; POF Photochemical ozone formation; RI Respiratory inorganics; HTNC Human toxicity, non-cancer effects; HTC Human toxicity, cancer effects; A Acidification; EAF Eutrophication, aquatic freshwater; EAM Eutrophication, aquatic marine; ET Eutrophication, terrestrial; EF Ecotoxicity freshwater; LU Land use; WU Water use; RUE Resource use, energy carriers; RUM Resource use, mineral, and metals

Finally, as regards the *Sensitivity scenario*, the differences between the systems are significantly more relevant. This is due to both the important increase of impacts of the bioplastic bag system under the worst operational conditions and the improvements introduced for the paper bag system. Thanks to the latter, the paper bag system is characterized by a result negative in sign with both modeling approaches.

Table 3 Life Cycle Assessment results: potential impacts of the *Baseline scenario* and the *Sensitivity scenario*, calculated with the Environmental Product Declaration (EPD) and the Product Environmental Footprint (PEF) approaches, for the paper bag system and for the bioplastic bag system (dedicated and shopper), after the normalization and weighing stages. Results are indicated in μPt per functional unit

Scenario	EPD approach			PEF approach		
	Paper bag system	Bioplastic bag system (dedicated)	Bioplastic bag system (shopper)	Paper bag system	Bioplastic bag system (dedicated)	Bioplastic bag system (shopper)
Baseline scenario	−0.46	2.19	0.72	0.81	1.89	0.53
Sensitivity scenario	−1.55	8.33	5.73	−0.28	8.24	5.70

4 Conclusions

According to the findings of the study, the use of paper bags for the food waste collection in the Italian system, currently widely based on bioplastic bags, should be encouraged. First, the reduction of the amount of waste to be collected and the lower odor and leachate release during the use of paper bags at the household could pave the way to a potential decrease of the food waste collection frequency, then reducing the environmental impacts. As regards the food waste treatment process, while bioplastic bags are discarded as residues, there is a very good compatibility of the paper bags with the AD process, leading to potential energy benefits: according to semi-continuous tests, the use of paper bags allows for an 8% increase in the methane production per mass unit of food waste in addition to that obtained from the sole food waste digestion. Finally, the use of paper bags allows for the reduction of the potential environmental impacts of the current food waste management based on bioplastic bags. Globally, paper bags can improve the current food waste management system allowing for a reduction of treatment residues and an improvement of the energy valorization leading to a reduction of the potential environmental impacts. Accordingly, the outcomes of the study can significantly contribute to the fulfillment of the selected SDGs.

Acknowledgements I would like to acknowledge my supervisor Prof. Mario Grosso for the help and the support in the development of this research project. Many thanks to the other Professors and colleagues that supported and made this work possible,

References

1. ISPRA—Istituto Superiore per la Protezione e la Ricerca Ambientale: Rapporto rifiuti urbani edizione 2020—Urban waste report (2020). https://www.isprambiente.gov.it/files2020/pubbli cazioni/rapporti/rapportorifiutiurbani_ed-2020_n-331-1.pdf. Accessed 14 Oct. 2021
2. CIC-COREPLA: Studio CIC-COREPLA, 2020: triplicano le bioplastiche compostabili nella raccolta dell'organico—Compostable bioplastics in the food waste collection tripled (2020). https://www.compost.it/wp-content/uploads/2020/07/CS-4-Studio-CIC-COREPLA-2020_-raddoppiano-le-bioplastiche-compostabili-nella-raccolta-dell%E2%80%99organico-5.pdf. Accessed 4 Apr. 2022
3. Utilitalia: managing and recovering bioplastics. Utilitalia Position Paper adopted on January 21 2020 by the Environmental Board (2020). http://www.utilitalia.it/dms/file/open/?8bf343a1-901d-4f99-890b-bbe422a324fd. Accessed 14 Oct. 2021
4. UNI EN: UNI EN 13432:2002 Packaging—Requirements for packaging recoverable through composting and biodegradation—Test scheme and evaluation criteria for the final acceptance of packaging (2002)
5. Fazio, S., Castellani, V., Sala, S., Schau, E.M., Secchi, M., Zampori L.: Supporting information to the characterisation factors of recommended EF life cycle impact assessment methods. EUR 28888 EN. European Commission, Ispra (2018). https://doi.org/10.2760/671368
6. Ecoinvent centre: ecoinvent version 3.5 database.: (2018). http://www.ecoinvent.org/. Accessed 14 Oct 2021
7. EPD International: General Programme Instructions for the International EPD System. Version 3.01 (2019). www.environdec.com. Accessed 14 Oct 2021
8. EC—European Commission: Product Environmental Footprint Category Rules (PEFCR) Guidance document, Guidance for the development of Product Environmental Footprint Category Rules (PEFCRs), version 6.3 (2018). https://ec.europa.eu/environment/eussd/smgp/pdf/PEFCR_guidance_v6.3.pdf. Accessed 14 Oct. 2021
9. EC: Recommendation 2013/179/EU of 9 April 2013 on the use of common methods to measure and communicate the life cycle environmental performance of products and organisations. Official J. Euro. Union L **124**, 4 (2013)

Flood Damage Assessment to Economic Activities in the Italian Context

Marta Galliani

Abstract In the last century the number of floods affecting people increased across Europe, due to both more frequent intense events and the growth of population and urbanization in flood-prone areas. Equipping cities with tools for flood damage assessment is crucial to effectively manage and reduce flood risk. The sector of businesses has a key role in cities development and suffers high losses in case of inundation, but damage appraisal to economic activities is still a challenging task. This study took up the challenge of addressing this topic, with specific reference to direct damage and the Italian context. Two approaches have been implemented: the analysis of about a thousand damage data regarding economic activities in four Italian flood events and the development of damage functions for retail activities by means a synthetic approach. The results led to the identification of the most vulnerable elements of different types of economic activities and provided reference values to assess the order of magnitude of flood damage.

Graphical Abstract

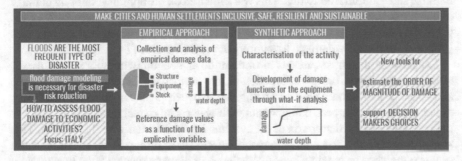

Keywords SDG 11 · Flood · Damage · Flood risk · Economic activities · Damage model

M. Galliani (✉)
Department of Civil and Environmental Engineering (DICA), Politecnico di Milano, Piazza Leonardo da Vinci 32, 20133 Milano, Italy
e-mail: marta.galliani@polimi.it

1 Introduction

In the last twenty years, flooding was the most frequent type of natural disaster, affecting more than one and half billion of people and causing damage equal to 651 billion US$ in the world [1]. Nonetheless, flood damage is expected to further increase in the future, as a consequence of more frequent extreme events caused by climate change [2–5] as well as of the new urbanization of flood-prone areas, leading to growing exposure of people and resources [6, 7]. The implementation of suitable Disaster Risk Reduction (DRR) measures is key for reducing the impacts of extreme events [8]. The Sendai Framework for Disaster Risk Reduction [9], adopted at the Third UN World Conference in Sendai on March 2015, recognizes the need of developing and implementing DRR policies as a priority at the international level. In detail, it identifies four priorities of intervention: (i) understanding disaster risk, (ii) strengthening disaster risk governance to manage disaster risk, (iii) investing in disaster risk reduction for resilience, and (iv) enhancing disaster preparedness for effective response and to "Build Back Better". The 2030 Agenda for Sustainable Development [10] reaffirms the objectives of the Sendai Framework and the need of reducing disaster risk within the Goal 11: "make cities and human settlements inclusive, safe, resilient and sustainable". This Goal proposes to significantly reduce economic losses caused by disasters and to increase the number of cities developing and implementing holistic disaster risk management at all levels. In order to achieve these objectives and guarantee an efficient allocation of financial resources for risk mitigation, the quantification of potential damage caused by disastrous events like floods is required [11].

According to the EU Floods Directive 2007 [12], flood damage is the potential adverse consequences of floods for human health, the environment, the cultural heritage and economic activities. This work focused on the appraisal of flood damage to economic activities, in the Italian context. Indeed, although in Italy 24% of the territory is exposed to flood risk, including 18% of industrial and service premises [13], tools to quantify damage to enterprises are still scarce [14, 15].

Quantifying the consequences of flooding to an economic activity is not an easy task, as they depend on the interaction among numerous factors acting over time and space. For instance, we can distinguish between physical damage, due to the direct contact of premises components with water, and damage related to business interruption or to the disruption of the systems interconnected with the activity. This study focused on the first type of damage, referred to as direct damage [16]. The aim of this research is to obtain a better knowledge of damage mechanisms to economic activities and develop tools to represent and quantify direct damage to this sector in flooding events. Two approaches were implemented to reach this goal, which are defined as empirical and synthetic approach.

The first consisted in the analysis of empirical damage data, collected in the aftermath of flood events occurred in Italy, aiming at the acquisition of better knowledge on types and magnitude of damage to economic activities in case of flood. Results supply a first estimate, although rough, of flood damage to economic activities in Italy in relation to its main explanatory variables: water depth, activity surface and activity type.

The second approach aimed at developing damage functions by means of an expert analysis (what-if questions) of damage mechanisms [17]. The synthetic analysis focused only on retails trade activities and restaurants and implied a detailed characterization of the equipment of the activities.

2 Methods

2.1 Empirical Approach

The empirical approach was implemented on a dataset including post-event damage records for four case studies:

- *Lodi*. The flood occurred in the town of Lodi (Northern Italy) in November 2002 due to the overflow of the Adda river [18, 19].
- *Secchia*. The flood occurred in the province of Modena (Northern Italy), in January 2014 [20], caused by a dike breach along the river Secchia. Data refer to three municipalities: Bastiglia, Bomporto, and Modena.
- *Enza*. The bank breakage of Enza River in the municipality of Brescello (Northern Italy) caused the flooding of the village Lentigione in December 2017 [21].
- *Sardegna*. The floods due heavy rain flows and bad weather in Sardegna Region (Southern Italy) in November 2013. Collected data refer in particular to the city of Olbia (Northern Sardegna).

For each affected activity, the dataset contains information about the water depth, the damage, and the characteristics of the activity. Information on water depth was obtained from hydraulic modeling [18, 20]. Damage data derive instead from the declarations filled in by the owners of the affected enterprises, to ask for national compensations. Damage data is specified for three components: damage to structure, equipment, and stock, where structure identifies the building with the internal systems necessary to the function of the building (e.g., electrical system or heating system); equipment refers to machineries, furniture, vehicles, and tools necessary for functions of business; stock refers to raw material, semi-finished and finished products. The information about the activities differs according to the case study, as it depends on the degree at which authorities processed the claims and on the actual information included in the original forms. For instance, for all case studies, the information about the type of activity (identified by the NACE code, that is Statistical Classification of Economic Activities adopted in the European Community [22]) is present. Differently, information about the number of employees is available only in Secchia claims. Information about the area of the activity, if not present in the claims, was

Table 1 Information about the cases studies and number of activities per provided information

Case study	Event date	Source		Information on affected activities			
			Damage	Water depth	Activity type	N employees	Surface
Lodi	Nov-02	Municipality	88	77	87	–	83
Sardegna	Oct-13	Region	637	240	514	–	431
Secchia	Jan-14	Region	226	226	201	105	142
Enza	Dec-17	Region	46	46	42	–	46

computed through GIS based-tools as footprint area of the building in which the activity is located. The main information included in the claims is summarized in Table 1

The dataset was used to implement various analyses aiming at (i) identifying the composition of damage for different activity categories, (ii) studying the relation of damage with its explanatory variables, (iii) computing the relative damage. All the analyses were based on an essential conceptualization of damage: each component of damage $D_{component}$ (i.e. structure, equipment and stock) was expressed as a function of three significant variables, that are activity type, activity surface, and water depth (Eq. 1).

$$D_{component} = f(\text{activitytype, waterdepth, activitysize}) \tag{1}$$

In fact, a first analysis investigated the composition of damage according to the activity type. The information about the activity type, identified by the NACE code, was joined with information about damage components, in order to observe if there were similar behaviors in the case studies and to obtain the average composition of damage. This analysis focused only to the NACE categories with more than 10 data and considered representative of the sector "economic activities" (e.g., no agriculture or infrastructure). In detail, the NACE codes were aggregated into four macro-categories, analyzed in this study, being: "Manufacturing" (NACE C), "Commercial" (NACE G), "Restaurant" (NACE I) and "Office" (NACE J, K, L, M, N).

A second analysis studied the relation of damage with water depth splitting data per damage component, activity categories (Manufacturing, Commercial, Restaurant and Office) and classes of water level. Furthermore, the relation of damage with activity size, related to the footprint area of the activity, was investigated, for different activity types; the average damage was computed for surface intervals and for the macro-categories of activities, to observe the prevalent trend. For these analyses, damage data were not divided per case study but revaluated to the year of the most recent event (2017), by considering the variation in the harmonized consumer price index supplied by ISTAT.

The empirical approach also aimed at studying the relation between the observed damage and the exposed value of the activity, in order to evaluate a relative damage. To compute the exposed values, a simplified method, based on the Flood-IMPAT procedure, was implemented [14]. The procedure considers the net capital stock as a measure of the exposed value of an enterprise. The net capital stock is the sum of the value of buildings, machinery, equipment, cultivated biological resources and intellectual property products of activities, and is provided, aggregated for NACE classes, by ISTAT. To estimate the exposed value, only the value components referring to structure and equipment were considered, for each NACE category. The value of the stock was instead not evaluated, as stock is not considered in the definition of the net capital stock, and no other sources of information were found for its estimation. Indeed, the stock value is hard to appraise due to the variability of goods constituting the stock, the variability of costs of these goods, and the variability of the amount of stock in time. The value of structure and equipment for NACE category was then divided by the total number of units (at the national level) per NACE class, to obtain a unitary reference value (e.g. the net capital stock per unit). Once the exposed value was estimated, the relative damage was computed per activity category as the ratio between the observed damage and the product between the exposed value and the number of activities of the same category.

The whole empirical analysis allowed to compute reference damage values as a function of different variables (i.e. activity type, activity surface, water depth). To guarantee the usability of results, such reference values were computed for various implementation scenarios, characterized by different available information on which performing the estimation. Consistency of results was verified comparing simulated and observed damage in the case studies of Lodi, Secchia, Sardegna and Enza.

2.2   Synthetic Approach

The synthetic analysis focused on the development of a new set of synthetic stage-damage functions for the assessment of damage to equipment of some types of retail trade activities. The development of the damage functions was the final step of a process of characterization and classification of economic activities developed in the project Flood-IMPAT+ [18] (www.floodimpatproject.polimi.it). Different types of retail trade activities were characterized in terms of typical size, main equipment components and reference costs to estimate the exposed value of equipment. Information was collected from national regulations, handbooks, AutoCAD libraries, commercial design sites, furniture and equipment catalogues, estimates for shop fittings. Table 2 shows, for example, the characterization of a pharmacy. Then, activities were classified in clusters on the basis of commonalities in equipment components (and in their vulnerability) and then in expected damage mechanisms (Table 3). The sum of the costs of the equipment elements constitutes the total value of the equipment, in terms of maximum, minimum and average value. Dividing the total value by the dimension, the equipment value as $€/m^2$ is obtained.

To develop damage curves, three further steps were implemented. First, for each equipment element, the water level for which the element is damaged was assumed, according to the nature of the element. Electric appliances (as counter, refrigerator,

Table 2 Characterization of pharmacy type

Pharmacy | Area 60 m^2

Element	Elevation (cm)	Height (cm)	Quantity	Cost of single element [€]		
				Min	Max	Average
Gondola shelving	20	120	4	738	1,563	1,150.5
Display furniture	20	200	14	295	477	386
Counter	0	100	3	960	960	960
Anti-shoplifting kit	0	150	4	1,000	1,700	1350
Desk	0	74	1	225	225	225
Chair	0	90	2	75	75	75
Galenical Laboratory	0	75	1	1,250	3,400	2,325
Drawer cabinets for medicines	0	220	8	1,376	1,930	1,653
Refrigerators	5	180	1	1,780	5,700	3,740
Warehouse shelves	10	200	8	525	894	709.5
Equipment value				32,575	54,677	43,626

Table 3 Clusters of non-food retail trade activities

Clusters of non-food retail activities	
1	Clothing, footwear, underwear, leather goods, shirts, costume, sporting goods
2	Pharmacy, herbalist's shop, medical and orthopedic articles, optics
3	Tobacconist, stationery, receipt-lot, newsstand, bookshop, comics
4	Household shop, soaps, gifts, appliances, electronics, informatics, toys, telephony, photography
5	Hardware store, paint factory, building materials, sanitary articles, gardening, security systems
6	Jeweler, silverware, watchmaking
7	Art objects, art gallery, articles for fine arts, religious articles, philately
8	Pet shops, aquariums, florists
9	Musical instruments
10	Funeral and cemetery items
11	Cars, motorcycles, vessels, bicycles
12	Service stations/petrol pumps

anti-shoplifting systems), wood elements and equipment for cooking were considered damaged at the minimum water level reaching the element, considering their elevation from the floor level. Metal elements, as shelves or cabinets, were considered damaged when water depth reach around half the height of the element. Second, the replacement costs of the elements damaged at different water level were added up to obtain the total absolute damage for each water depth. Third, absolute damage was divided by the sum of values of all elements (the equipment value) to obtain the relative damage. It is worth noting that replacement is the only considered cost, thus the only considered damage. Costs related to process as repair or cleaning were not included. The damage functions were finally tested using data collected for the empirical approach: Lodi, Sardegna, Secchia and Enza. These data had the necessary information to implement the damage functions, i.e., activity type, water depth and activity surface. The latter is required to evaluate the total value of the equipment of the activity, having previously calculated the value as €/m^2 for the types of activities studied. Because observations refer to different years than prices assumed to evaluate equipment value (2019), observed damage were discounted to the 2019 price value.

3 Results

3.1 Empirical Approach

The first result of the empirical analysis was the characterization of damage composition for the activity categories Manufacturing, Commercial, Restaurant and Office. Similarities were observed among the case studies, thus the fraction of damage components, on the total damage, was computed per activity type as weighted average on the number of data per case study (Table 4). Secondly, the analysis of the relation of damage with its explanatory variables and the computation of the relative damage lead to the results shown in Table 5.

Table 5 shows reference damage values obtained from the analysis, as functions of different variables. Such values can be used for a first estimation of damage, according to the available information for the assessment. For example, the first row corresponds to the case in which no information is available except the fact that the activity is flooded. In this case, the reference damage value was computed as average of total damage of the affected economic activities (excluding agriculture and infrastructure) and could be used as first, rough damage estimation. The second row supplies the average damage as €/m^2, in the case at least information on the activity size is available. The reference values were computed as ratio between the total damage and the sum of surface of all the affected activities. The third scenario

Table 4 Fraction of damage component, i.e. structure, equipment, and stock, on total damage	Fraction of damage component on the total			
	Activity category	Structure	Equipment	Stock
	Manufacturing	0.20	0.44	0.36
	Commercial	0.25	0.27	0.48
	Restaurant	0.46	0.44	0.10
	Office	0.63	0.30	0.07

Table 5 Reference damage values to a single economic activity, per different scenarios of available knowledge

Scenario of available knowledge		Unit of measure		Average damage	
1	No information	(€/unit)		Total damage	
				72000	
2	Activity surface	(€/m²)		Total damage	
				66	
3	Activity surface Water depth	(€/m²)	Water depth (m)	Total	
			0.0–0.3	40	
			0.3–0.6	70	
			0.6–1.0	90	
			1.0–1.5	95	
4	Activity surface Activity type	(€/m²)	Activity type	Total damage	
			Manufacturing	70	
			Commercial	85	
			Restaurant	120	
			Office	30	
5	Activity type Exposed value of structure and equipment	Relative damage (/)	Activity type	Structure	Equipment
			Manufacturing	0.08	0.10
			Commercial	0.13	0.30
			Restaurant	0.05	0.37
			Office	0.07	0.10

corresponds to the case in which both the surface of the activity and the water depth at its location are known. The analysis of the relation of damage with water depth revealed an increasing trend of total damage with water depth. Still, the same trend was not visible considering the activity type. In fact, dividing the dataset by type of activity considerably reduces the number of data for the analysis, to the point they are not sufficient to observe a representative trend. Thus, the reference damage values are supplied expressed in €/m² per water depth intervals but not per activity type. In the fourth scenario the activity type is supposed to be known and the damage values are provided as a function of the surface. The table supplies only the total damage, but the damage components can be computed knowing the damage composition in Table 4. The reference values were computed as ratio between the total damage and the sum of surface of the affected activities per activity category. In the last scenario, the available information is the type of the activity and the exposed value. The reference damage value is the relative damage per activity category and component, to be multiplied by the exposed values to get an estimate of the absolute damage. Relative damage is supplied only for structure and equipment, because of the lack of data/method to compute the exposed value of the stock (see Sect. 2.1). However, damage to stock can be calculated knowing the portion of damage to stock in the total (Table 4).

Consistency of results (i.e., reference damage values) obtained for the various scenarios of available knowledge was verified by comparing simulated and observed damage for the four case studies (Table 6). The comparison was performed only for the activities belonging to the investigated categories and with information about water depth and surface.

3.2 Synthetic Approach

The result of the synthetic analysis is a set of stage-damage functions for the types of activities: clothing shop, pharmacy, tobacconist, supermarket, and restaurant. To be noted that the first three functions can be considered valid for all the types of activities included in the respective clusters (Table 3). Figure 1 shows the functions in terms of relative damage. To obtain the functions in absolute terms, it is necessary to multiply the relative damage by the equipment value. The characterization of these activities provided the computation of equipment value, as sum of values of the single elements that compose the equipment (Sect. 2.2). Table 7 shows the average equipment value for the analyzed activity types.

Damage simulated by these functions was compared to observed damage for a small set of activities, composed by the enterprises of Lodi, Secchia, Sardegna and Enza that belong to the typologies clothing shop, pharmacy, tobacconist, supermarket and restaurants. Table 7 compares observed and estimated damage in the four case studies, in terms of absolute damage, for the activities of interest.

Table 6 Comparison between observed and computed damage with reference damage values supplied in Table 5

		N data	Observed damage (2017)	Scenario of available knowledge				
				1	2	3	4	5
Total damage 10^6 €	Lodi	56	2.2	4.0	3.2	3.4	3.7	4.0
	Sardegna	112	4.0	8.1	2.1	2.5	2.3	6.1
	Secchia	90	8.0	6.5	8.0	9.2	8.2	8.0
	Enza	20	5.2	1.4	5.1	6.5	4.7	1.5
Computed/observed	Lodi			1.8	1.4	1.5	1.6	1.8
	Sardegna			2.0	0.5	0.6	0.6	1.5
	Secchia			0.8	1.0	1.1	1.0	1.0
	Enza			0.3	1.0	1.2	0.9	0.3

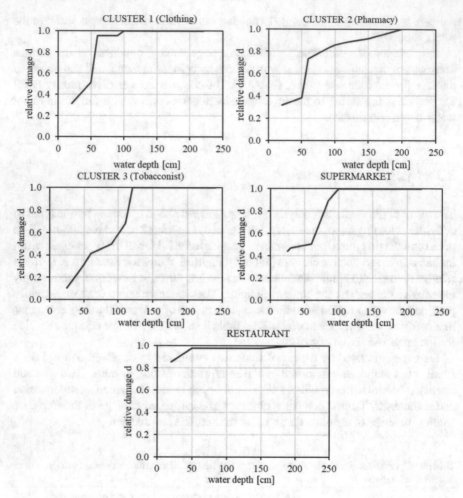

Fig. 1 Stage-damage curves of equipment of cluster 1 (clothing), cluster 2 (pharmacy), cluster 3 (tobacconist), supermarket, and restaurant activities

4 Discussion and Conclusion

The results of both approaches constitute a base knowledge to develop a more complete damage modeling tool. In fact, the empirical approach supplies reference damage values that can be used to assess the damage for macro-categories of activities as a function of different variables. The synthetic approach provides simple damage functions to assess damage to equipment for a limited number of activity types, but more specific, and as function of water depth only. These tools can be used together, according to the available information, to obtain an appraisal of the order of magnitude of potential damage.

Table 7 Comparison of observed and calculated damage using the equipment damage functions in Fig. 1

Clusters and activity types		Equipment value (€)	n° activities	Sum observed damage (10^3 €)	Sum calculated damage (10^3 €)	Calculated damage/observed damage
1	Clothing	13,070	10	54	257	4.8
2	Pharmacy	43,626	2	211	129	0.6
3	Tobacconist	12,854	8	24	755	31.9
	Supermarket	66,180	12	120	457	3.8
	Restaurant	49,120	23	406	318	0.8
	Weighted average					6.7

Despite these results represent a first attempt of developing a flood damage model for economic activities in Italy, the error computed from the comparison with observed damage has the same order of magnitude of more developed models. In the synthetic approach, excluding tobacconists, simulated/observed ratio varies between 0.6 and 4.8 (Table 7). These results do not exceed expectations and are in line with the estimation errors observed in other case studies. For instance, for the residential sector, analysis in [19] revealed an average ratio between damage calculated by different European models and observed damage of 4.06. Reasons of estimation errors are related not only to the uncertainty of the model, but also to the quality of data, deriving from different citizens and processed from different authorities.

To conclude, this study contributes to improve the present knowledge on damage mechanisms to economic activities in the Italian context. In particular, the performed analyses led to the characterization of damage composition, and to the estimate of relative damage and mean absolute damage by categories of activity. The study faced some obstacles as the little number of data, the lack of homogeneity in the information included in collected damage data and the lack of available and complete databases to characterize the enterprises. The results need to be tested and validated with further data and in new case studies, before actually be delivered to end-users. Nevertheless, they allow to obtain a quantitative appraisal of potential damage, that is recommended for an objective and transparent evaluation of protection measures, both before and after a disastrous event. Moreover, quantitative assessment of damage is at the base of cost–benefit analyses implemented to evaluate effectiveness of mitigation interventions and their prioritizing. It is also useful for managing damage compensation in the post-event, especially in countries like Italy, where the insurance system is not enough diffused and compensation is mostly in charge of public authorities.

The current global context, characterized by an increase of intense events and risk-prone assets, requires an increasing effort in actions of prevention and mitigation of risk. Flood damage assessment to economic activities contributes to improve these actions, considering the central role of businesses in the wellbeing of society [11].

Consequently, it contributes to the management of more safe and resilient cities and to the achievement of goals of 2030 Agenda for Sustainable Development.

Acknowledgements The author acknowledges with gratitude the professors Daniela Molinari and Francesco Ballio (Politecnico di Milano) for their contributes and valuable support in this work and Giulia Pesaro, Scira Menoni and Guido Minucci (Politecnico di Milano) for their first analysis of Lodi data and their fruitful suggestion during the development of the work.

References

1. CRED Centre for Research on the Epidemiology of Disasters, UNDRR United Nations Office for Disaster Risk Reduction: The human cost of disasters: an overview of the last 20 years (2000–2019). https://www.undrr.org/publication/human-cost-disasters-overview-last-20-years-2000-2019. Accessed 4 Oct 2021 (2020)
2. World Bank Group, Guide to Climate Change Adaptation in Cities, World Bank, Washington, DC. © World Bank. https://openknowledge.worldbank.org/handle/10986/27396 (2011)
3. Govind, P.J., Verchick, R.R.M.: Natural disaster and climate change. In: International Environmental Law and the Global South, pp. 491–507. Cambridge University Press. https://doi.org/10.1017/CBO9781107295414.024, (2015).
4. Alfieri, L., Bisselink, B., Dottori, F., Naumann, G., de Roo, A., Salamon, P., Wyser, K., Feyen, L.: Global projections of river flood risk in a warmer world. Earth's Fut. **5**, 171–182 (2017). https://doi.org/10.1002/2016EF000485
5. Prein, A.F., Rasmussen, R.M., Ikeda, K., Liu, C., Clark, M.P., Holland, G.J.: The future intensification of hourly precipitation extremes. Nat. Clim. Chang. **7**, 48–52 (2017). https://doi.org/10.1038/nclimate3168
6. Barredo, J.I.: Major flood disasters in Europe: 1950–2005. Nat. Hazards **42**, 125–148, https://doi.org/10.1007/s11069-006-9065-2 (2007)
7. Paprotny, D., Sebastian, A., Morales-Nápoles, O., Jonkman, S.N.: Trends in flood losses in Europe over the past 150 years. Nat. Commun. **9**, 1985 (2018). https://doi.org/10.1038/s41467-018-04253-1
8. Ward P.J., de Ruiter M. C., Mård J., Schröter K., Van Loon A., Veldkamp T., von Uexkull N., Wanders N., AghaKouchak A, Arnbjerg-Nielsen K., Capewell L., Llasat M.C., Day R., Dewals B., Di Baldassarre G., Huning L.S., Kreibich H., Mazzoleni M., Savelli E., Teutschbein C., van den Berg H., van der Heijden A., Vincken J.M.R., Waterloo M.J., Wens M.: The need to integrate flood and drought disaster risk reduction strategies. Water Secur. **11** (2020)
9. UNDDR United Nations Office for Disaster Risk Reduction: Sendai Framework for Disaster Risk Reduction 2015–2030, Geneva, Switzerland, available at https://www.undrr.org/publication/sendai-framework-disaster-risk-reduction-2015-2030. Accessed 4 Oct 2021 (2015)
10. UN United Nations General Assembly: Transforming our world: the 2030 Agenda for Sustainable Development. https://sdgs.un.org/2030agenda. Accessed 4 Oct 2021 (2015)
11. Menoni, S., Molinari, D., Ballio, F., Minucci, G., Mejri, O., Atun, F., Berni, N., Pandolfo, C.: Flood damage: a model for consistent, complete and multipurpose scenarios. Nat. Hazards Earth Syst. Sci. **16**, 2783–2797 (2016). https://doi.org/10.5194/nhess-16-2783-2016
12. European Parliament and the Council of the European Union: Directive on the assessment and management of flood risks (2007/60/EU). Off. J. L **288** (2007)
13. Trigila, A., Iadanza, C.: Landslides and floods in Italy: hazard and risk indicators. Summary Report 2018, ISPRA, Dipartimento per il Servizio Geologico d'Italia—Geological Survey of Italy (2018)
14. Molinari, D., Minucci, G., Mendoza, M.T., Simonelli, T.: Implementing the European "Floods Directive": the Case of the Po River Basin. Water Resour. Manag.: Int. J. Publ. Euro. Water

Resour. Assoc. (EWRA). Springer; European Water Resources Association (EWRA), **30**(5), 1739–1756 (2016)

15. Marin, G., Modica, M.: Socio-economic exposure to natural disasters. Environ. Impact Assess. Rev. **64**, 57–66. ISSN 0195-9255, https://doi.org/10.1016/j.eiar.2017.03.002 (2017)
16. Merz, B., Kreibich, H., Schwarze, R., Thieken, A.: Review article "Assessment of economic flood damage." Nat. Hazards Earth Syst. Sci. **10**, 1697–1724 (2010). https://doi.org/10.5194/nhess-10-1697-2010
17. Smith, D.I.: Flood damage estimation—A review of urban stage-damage curves and loss functions. Water SA **20**(3), 231–238 (1994)
18. Molinari, D., Minucci G., Gallazzi A., Galliani M., Mendoza M.T., Pesaro G., Radice A., Scorzini A.R., Menoni S., Ballio F.: Del.3: Strumenti per la modellazione del danno alluvionale, deliverable of the project Flood-IMPAT+ an Integrated Meso & Micro Scale Procedure to Assess Territorial Flood Risk. www.floodimpatproject.polimi.it (2019)
19. Molinari, D., Scorzini, A.R., Arrighi, C., Carisi, F., Castelli, F., Domeneghetti, A., Gallazzi, A., Galliani, M., Grelot, F., Kellermann, P., Kreibich, H., Mohor, G.S., Mosimann, M., Natho, S., Richert, C., Schroeter, K., Thieken, A.H., Zischg, A.P., Ballio, F.: Are flood damage models converging to "reality"? Lessons learnt from a blind test. Nat. Hazards Earth Syst. Sci. **20**, 2997–3017 (2020). https://doi.org/10.5194/nhess-20-2997-2020
20. Carisi, F., Schröter, K., Domeneghetti, A., Kreibich, H., Castellarin, A.: Development and assessment of uni- and multivariable flood loss models for Emilia-Romagna (Italy). Nat. Hazards Earth Syst. Sci. **18**, 2057–2079 (2018). https://doi.org/10.5194/nhess-18-2057-2018
21. Regione Emilia-Romagna: Eccezionali eventi meteorologici che si sono verificati dall'8 al 12 dicembre 2017 nel territorio delle province di Piacenza, di Parma, di Reggio Emilia, di Modena, di Bologna e di Forlì-Cesena (OCDPC n. 503/2018. Approvazione del Piano dei primi interventi urgenti di protezione civile—Primo stralcio), Bollettino ufficiale della Regione Emilia-Romagna-Parte seconda, n. 103 (2018)
22. Eurostat, European Commission: NACE Rev. 2 Statistical classification of economic activities in the European Community. https://ec.europa.eu/eurostat/web/products-manuals-and-guidel ines/-/ks-ra-07-015 (2008)

Cold Atom Interferometry in Satellite Geodesy for Sustainable Environmental Management

Khulan Batsukh

Abstract Our Earth is a complex system. By monitoring the integrated geodetic-geodynamic processes, we can understand its sub-systems and geographical distribution of its resources. With the development of space techniques and artificial satellites, satellite geodesy era started, e.g., it became possible to observe a wide range of processes, occurring both on and below the Earth's surface. Such observations can be exploited not only in environmental activities, but also in societal activities like natural disasters monitoring. Thus, satellite geodesy can bring great benefits to "Climate action", one of the 17 sustainable development goals of the United Nation: we can estimate the ice-sheet mass balance and study the impact of climate change by monitoring sea levels. This paper aims to investigate the possible implementation of cold atom sensors for future satellite gravity missions, which would improve our current knowledge of the Earth's gravity field and contribute into the sustainable environmental management.

Graphical Abstract

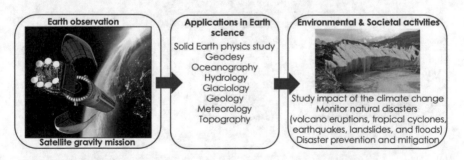

Keywords Earth observation · Satellite geodesy · Gradiometry · Cold atom interferometer · Climate action · Sustainable environmental management

K. Batsukh (✉)
Department of Civil and Environmental Engineering (DICA), Politecnico di Milano, Piazza Leonardo da Vinci 32, 20133 Milan, Italy
e-mail: khulan.batsukh@polimi.it

M. Antonelli and G. Della Vecchia (eds.), *Civil and Environmental Engineering for the Sustainable Development Goals*, PoliMI SpringerBriefs, https://doi.org/10.1007/978-3-030-99593-5_4

1 Introduction

The Earth's gravity field is the result of mass distribution in the Earth's system, which includes the solid Earth, oceans, atmosphere, ice, land, as well as the biosphere. Mass redistribution and transport in any of the Earth's subsystems cause gravity field variations. Thus, continuous monitoring of the Earth's gravity and its changes is important for studies on climate changes, hydrology, sea level changes, solid Earth phenomena (Fig. 1). These variations can be measured both on the Earth surface and from the space. Data from gravity surveys on the Earth surface allow us to map the gravity field with high resolution, but these data are inhomogeneous and have a different level of precision depending on the data availability and used instruments. Such issue can be overcome by using satellite-based observations which provide a uniform coverage of the Earth and map its gravity field in a very homogeneous way.

The aim of satellite gravity missions is to refine the Earth's gravity field model, monitor and model the static and/or the time-variable gravity field variation. The static gravity field represents the spatial variations in the gravity field, generated by the density inhomogeneities inside Earth, representing the geological processes acted over the geological time scale. Such observations can contain information on the presence of sediment basins, magmatic intrusions, volcanic deposits, metamorphic rock lineaments. At larger scale, they represent density contrasts associated with

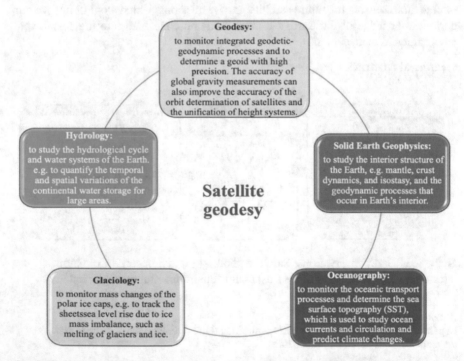

Fig. 1 Contribution of satellite geodesy in the Earth observation. Adapted from [14]

mantle convection or plate tectonics, tied to long term dynamical processes within the Earth. On the other hand, the time-variable gravity field is the expression of the variations that occur much more rapidly due to various tidal, oceanic, atmospheric, hydrological, glaciological, and tectonic processes. Measurement of such temporal variations can reveal the effect of mega earthquakes and monitor crustal deformation. Besides, ice masses carry a significant gravity signal and time-variable gravity solutions are fundamental for studies on ice-sheet mass variations and their impact on climate [25].

In the first decade of the new millennium, three successful satellite missions (Fig. 2) were launched:

1. CHAMP (Challenging Mini-satellite Payload): to observe both the gravity field and magnetic field from a low altitude orbit [28].
2. GRACE (Gravity Recovery and Climate Experiment): to map the variations of Earth's gravity by measuring the changes in the inter-satellite distance [31].
3. GOCE (Gravity field and steady Ocean Circulation Explorer): to determine the static gravity field: geoid, the equipotential reference surface, and gravity anomalies with unprecedented accuracy and spatial resolution [9].

These missions delivered plenty of data to the scientific communities and space agencies and, of course, achieved spectacular science results, but our current knowledge of the Earth's system is still limited. Thus, the sustained observation of the Earth's gravity field is required to overcome this shortcoming. The geophysical and societal impact and objectives of future gravity missions with respect to GRACE and GOCE have been studied, e.g., in the framework of the Global Geodetic Observing

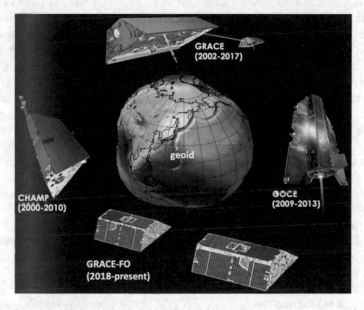

Fig. 2 Dedicated satellite gravity missions. Adapted from [12]

System (GGOS) Working Group for Satellite Missions [25]. Recently, a GRACE-FO (Gravity Recovery and Climate Experiment Follow-On) mission was launched in 2018 [15] (Fig. 2). It is continuing and will continue the work of previous GRACE twin in monitoring the water movement around the planet to detect the changes in amount of water in large lakes and rivers, soil moisture, ice sheets and glaciers, and in sea level caused by the water transfer from continental areas to the ocean [10].

Currently, ongoing studies such as NGGM (Next Generation Gravity Mission) are also being carried out for the long-term monitoring of the Earth's gravity field with high temporal and spatial resolution [13]. One of recent NGGM studies investigated its capability to detect the earthquake gravity signatures, including those from active tectonics and inter-seismic deformation [3]. It was proved that this new application of NGGM can be used to also detect the gravity signatures of those processes responsible for the earthquake generation, thus contributing to seismic hazard studies.

2 Spaceborne Cold Atom Gradiometer for Future Gravity Mission

Quantum sensors based on atom interferometry have evolved rapidly in the last few decades. Thanks to their excellent long-term stability and accuracy level, such sensors have demonstrated the potential to enable the inertial and gravity sensing payloads and have been used for a wide range of practical applications from oil and minerals exploration to defense, from civil engineering to space [1]. Future improvements of the GOCE mission concept can be achieved by going beyond the technology of electrostatic gravity gradiometers and taking advantage of a new generation of quantum sensors. Carraz et al. [4] proposed the concept of a new spaceborne cold atom gradiometer for high-sensitivity measurements of all diagonal elements of the gravity gradient tensor and the full spacecraft angular velocity. A gradiometer in one given direction (e.g., the radial one) can be obtained by measuring the acceleration of two clouds of cold atoms (at mK) separated by a certain distance. At the same time, this instrument allows to measure the rotation angle around the axis perpendicular to the plane of motion of the atom clouds. Several studies of European Space Agency (ESA) [7, 23, 33] investigated the performance of an innovative gradiometer and technological developments required for further space applications. The power spectral density of the gradiometer noise is flat at low frequencies in contrast to the spectrally colored noise of the electrostatic accelerometers (Fig. 3). The sensitivity of the ground-based atom gradiometers can reach a few tens of $E/Hz^{1/2}$ [30], while a spaceborne cold atom gradiometer can provide the sensitivity of few $mE/Hz^{1/2}$ [4], where $1\ E = 10^{-9\,s-2}$. Thanks to its spectral characteristics, this instrument might allow to meet the requirements of a future mission dedicated to the observation of both the time-variable gravity field (like GRACE) and the static field (like GOCE).

One of possible schemes of a cold atom gradiometer has been proposed and implemented in the framework of the MOCASS (Mass Observation with Cold Atom

Fig. 3 Electrostatic (GOCE) and cold atom gradiometer (MOCASS and CAI) noise spectra

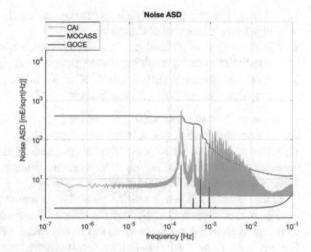

Sensors in Space) study, performed by Italian universities and research institutions under the Italian Space Agency (ASI) contract [22, 27, 29]. A similar concept also has been exploited in the "Cold Atom Inertial Sensors: Mission Application (CAI)" study, carried out by Thales Alenia Space Italia under ESA contract [23]. In the former, the gradiometer capability to detect and monitor various geophysical phenomena was assessed, considering different gradiometer configurations, whereas the latter focused more on the possible payload of a single-arm gradiometer and the mission architecture, addressing the required accommodation resources and technological developments. Other main differences between two studies were in the simulation scenarios, as well as in the characteristics of the gradiometer noise (see again Fig. 3).

3 Data Analysis

In both studies, the gradiometer performance was evaluated by using numerical simulations, processing the simulated data by space-wise approach. The main idea behind this approach [18] is to filter the noise and estimate the spherical harmonic coefficients of the geopotential model by exploiting the spatial correlation of the Earth's gravity field. Theoretically, it could be done by a global collocation solution, modelling the signal covariance as a function of spatial distance, and the noise covariance as a function of time distance. It would allow to filter together data that are close in space but far in time, thus overcoming the problems related to the strong time correlation of the observation noise. However, such a unique collocation solution is computationally unfeasible due to the huge amount of data to be processed for this type of satellite missions [24]. Indeed, the dimension of the system to be solved in this case would be equal to the number of input data. Hence, the space-wise approach (Fig. 5) was implemented as a multi-step collocation procedure consisting of:

1. Wiener filtering [26], applied to the data along the orbit to reduce the highly time correlated noise of the gradiometer.
2. Interpolation of filtered data to obtain values over spherical grids at mean satellite altitude, by applying collocation to local patches of data [20].
3. Spherical harmonic (SH) analysis of gridded data by numerical integration [5] to retrieve the geopotential coefficients.

In the case of the cold atom gradiometer, time correlation arises also from the atom interferometer integrator or transfer function [32], which is shown in Fig. 4a. This means that the MOCASS or CAI gradiometer does not provide point-wise observations like in GOCE and, therefore, the Wiener filter was generalized to a Wiener deconvolution filter (Fig. 4b).

In the general GOCE scheme, the procedure is iterated till convergence to recover the signal frequencies cancelled by the Wiener filter along the orbit and to correct the rotation from the gradiometer to the local orbital reference frame (LORF) [19]. For MOCASS and CAI simulations, no iterations were implemented due to the prevailing white behavior of the noise spectrum of the cold atom gradiometer and to the assumption that the spacecraft (and therefore the gradiometer) is kept aligned with the LORF. The simplified flow-chart in Fig. 5 without iterations allowed higher speed in computations and hence the possibility to analyze several case studies.

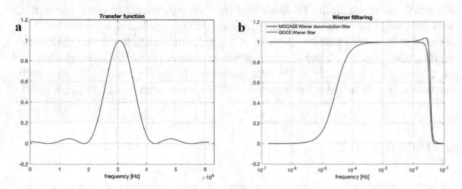

Fig. 4 Cold atom interferometer transfer function shape **a** and comparison of MOCASS Wiener deconvolution filter and GOCE Wiener filter **b** in the frequency domain

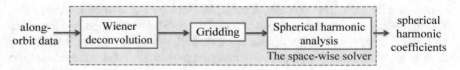

Fig. 5 A simplified flow-chart of the space-wise approach

Table 1 Simulation scenario in MOCASS and CAI studies

	MOCASS study	CAI study
Reference model	EIGEN-6C4	EIGEN-6C4
Reference orbit	2 months in high (259 km) and low orbit (239 km) from GOCE mission	2 months (259 km, GOCE-like)
Simulated data	Single-arm: T_{xx}, T_{yy}, T_{zz} Double-arm: T_{xx} and T_{zz}, T_{yy} and T_{zz}	Single-arm: T_{xx}, T_{yy}, T_{zz}
Mode of operation	Nadir-pointing mode (LORF) and inertial mode (IRF)	Nadir-pointing mode (LORF)
Number of MC samples	$N = 20$	$N = 50$

4 Simulation Scenario

The numerical simulations were performed considering the parameters reported in Table 1. In the case of MOCASS study, 20 different simulation scenarios were implemented, whereas only 3 scenarios were considered in CAI study. For all simulated scenarios, the estimation errors were computed using Monte Carlo (MC) techniques as an exact error covariance propagation is not feasible [21].

The MC strategy consisted in simulating a set of spherical harmonic model coefficients $T_{\ell \updownarrow}$ of the anomalous potential T of degree ℓ and order m, then computing average and sample error covariances of the corresponding solutions. The coefficient error variances of the reference model EIGEN-6C4 [11] were used to generate MC samples. This means that the sample data are equal to EIGEN-6C4 on average, but differ from it by a variation, related to the EIGEN-6C4 error variances. A higher number of samples would be recommendable to get more accurate error estimates; however, the choice was driven by the need of limiting the computational burden. The error rms of the estimated coefficients $\hat{T}_{\ell \updownarrow}$ was computed as:

$$\hat{\sigma}_{\ell m} = \sqrt{\frac{1}{N} \sum_{i}^{N} \left(\hat{T}_{\ell m}^{i} - T_{\ell m}^{i} \right)^2} \quad \forall \ell, m$$

5 Simulation Results

Let us consider 2-month MOCASS solutions in high orbit in terms of error degree variances (Fig. 6). For comparison, the error of the GOCE_TIM_R1 model [25] and the error of the average of two GRACE monthly solutions [17], corresponding to the same data period of the GOCE_TIM_R1 model, are reported here.

The best performance can be achieved when working with:

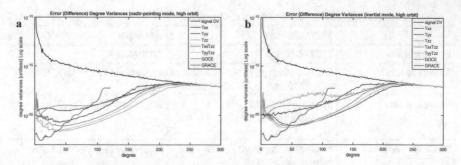

Fig. 6 MOCASS: Error degree variances in terms of estimated global gravity model: nadir-pointing mode (**a**) and inertial mode (**b**)

1. T_{zz} component or a pair of T_{yy} and T_{zz} in the case of a mission with a nadir-pointing satellite.
2. T_{yy} component or a pair of T_{yy} and T_{zz} in the case of a mission with an inertial-pointing satellite.

In these "optimal" scenarios, MOCASS observations allow for an improvement with respect to GOCE at all spherical harmonic degrees and with respect to GRACE at higher harmonic degrees (approximately above 35–40). At lower harmonic degrees, the GRACE estimates remain better than the MOCASS ones. Also, it must be underlined that this kind of comparison is a general indication of how well the MOCASS mission would perform with respect to GOCE and GRACE; in fact, the quality of real mission solutions depends on various factors such as the processing method or model version.

Consequently, the same comparison was done with CAI error curves, considering 2-month-solutions, which revealed that:

1. T_{zz} component shows the best performance, the error curves for T_{xx} and T_{yy} components show a quite similar performance.
2. In the "optimal" case, CAI shows improvement with respect to GOCE at degrees above 15, and with respect to GRACE at higher degrees (above 60).
3. Degradation of CAI with respect to MOCASS at all spherical harmonic degrees is due to the higher noise level of the CAI gradiometer, which in turn depends on the more accurate modelling of the interactions between the gradiometer and the spacecraft.

The results can also be compared in terms of gravity anomaly (Δg) cumulative error at ground level (Table 2):

1. Static gravity field: the commission error of the optimal MOCASS (in high orbit) and CAI solutions was computed at degree $\ell = 200$, which corresponds to spatial resolution of about 100 km. For a fair comparison, the corresponding cumulative error of the GOCE solutions (the time-wise solution of GOCE Release 1 (R1) and GOCE Release 5 (R5) [2]), is also reported.

Table 2 Commission error of the optimal MOCASS and CAI solutions

Solution	MOCASS (in high orbit)				CAI	GOCE TIM solution
	T_{zz} (LORF)	T_{yy} (IRF)	T_{yy} and T_{zz} (LORF)	T_{yy} and T_{zz} (IRF)	T_{zz}	
	Error in Δg at ground level [mGal] (at degree $\ell = 200$)					
Static gravity field (2-month solution)	1.03	1.24	1.84	1.67	2.00	2.91 (R1)
Static gravity field (5-year solution)	0.19	0.23	0.34	0.31	0.38	0.50 (R5)
	Trend error in Δg at ground level [μGal/month] (at degree $\ell = 45$)					GRACE solution
Time-variable gravity field (5-year mission with monthly solutions)	0.039	0.099	0.040	0.066	0.177	0.032

2. Time-variable gravity field: the commission error of the optimal MOCASS (in high orbit) and CAI solutions, including the GRACE one (ITSG-Grace2014k) [16], was computed at the maximum degree $\ell = 45$. It was due to the signal power, which is so lower that a higher accuracy is required for the signal detection, at the cost of a lower spatial resolution.

Here, the GOCE R1 solution corresponds to a time span of 2 months, whereas GOCE (TIM R5) and GRACE (ITSG-Grace2014k) solutions have been rescaled for the corresponding number of months.

6 Conclusion

Since 2014 the cold atom technology and instrument architecture have been further investigated for space application. So far, some achievements were obtained in operating atomic quantum sensors in the real-space environment, one of which is the Cold Atom Laboratory (CAL) [8]. It was successfully installed on the International Space Station in 2018 and serves as an experiment in the use of laser-cooled atoms for future quantum sensors. Recently, worldwide laboratories and research institutes have been investigating the idea of exploiting cold atom sensors for future satellite gravity missions [6]. Of course, designing such a mission is not only a technological challenge, but also represents complex trade-offs at the system and mission level.

The data simulation and analysis for two studies of future gravity missions, namely the MOCASS and CAI studies, were described in this paper. In both missions, the satellite would fly in a low orbit (GOCE-like) and carry onboard a gradiometer built by exploiting the ultra-cold atom technology. First estimates show that MOCASS and CAI gradiometer can improve the GOCE gradiometer performance in the mapping of the Earth's gravity field with high accuracy. Regarding the case of time-variable gravity field recovery, MOCASS and CAI error estimates show an improvement with respect to GRACE at higher degrees. Particularly, another fundamental part of MOCASS study was to evaluate the significance of the mission for the improvement in the detection and monitoring of geophysical phenomena, estimating the progress that could be achieved. For this purpose, different phenomena were simulated by geophysicists: deglaciation in High Mountains of Asia (HMA), mountain building processes (tectonic), continental hydrology, and volcanic eruptions leading to growing seamounts [27]. The simulation results showed that MOCASS observations could:

1. increase the detectability of small trends in the glacier ice mass in the HMA region and improve the studies of seamounts;
2. detect smaller areas subject to deglaciation in HMA;
3. map the complex tectonic deformation in the HMA with higher resolution.

Summarizing all the above mentioned aspects, it can be said that a cold atom interferometry is a promising maturing technology that could be implemented in future gravity missions to improve the measuring accuracy. A new gravity mission with such an innovative sensor may have great potential for enabling sustained observation of the Earth's gravity field and detecting and monitoring geophysical phenomena. For instance, by monitoring the sea levels, we can estimate the ice sheet mass balance and deglaciation rates and study the impact of climate change. Eventually, it may contribute into sustainable environmental management, and play a crucial role for achieving one of the sustainable development goals "Climate action".

Acknowledgements I would like to acknowledge my supervisors prof. Federica Migliaccio and prof. Mirko Reguzzoni for their continuous support and guidance throughout this work. The MOCASS study has been funded under Italian Space Agency contract N. 2016-9-U-0 "Proposal of a satellite mission and sensor concept based on advanced atom interferometry accelerometers for high resolution monitoring of mass variations on and below the Earth surface". The CAI study has been funded under European Space Agency contract N. 4000117930/16/NL/FF/gp.

References

1. Bongs, K., Holynski, M., Vovrosh, J., et al.: Taking atom interferometric quantum sensors from the laboratory to real-world applications. Nat. Rev. Phys. **3**, 814 (2021). https://doi.org/10.1038/s42254-021-00396-1
2. Brockmann, J.M., Zehentner, N., Höck, E., et al.: EGM_TIM_RL05: An independent geoid with centimeter accuracy purely based on the GOCE mission. Geophys. Res. Lett. **41**(22), 8089–8099 (2014). https://doi.org/10.1002/2014GL061904

3. Cambiotti, G., Douch, K., Cesare, S., et al.: On earthquake detectability by the next-generation gravity mission. Surv. Geophys. **41**, 1049–1074 (2020). https://doi.org/10.1007/s10712-020-09603-7
4. Carraz, O., Siemes, C., Massotti, L., et al.: A spaceborne gravity gradiometer concept based on cold atom interferometers for measuring Earth's gravity field. Microgravity Sci. Technol. **26**, 139–145 (2014). https://doi.org/10.1007/s12217-014-9385-x
5. Colombo, O.L.: Numerical methods for harmonic analysis on the sphere. Technical report N. 310, Department of Geodetic Science and Surveying, The Ohio State University, Columbus, Ohio (1981)
6. Devani, D., Maddox, S., Renshaw, R., et al.: Gravity sensing: cold atom trap onboard a 6U CubeSat. CEAS Space **12**, 539–549 (2020). https://doi.org/10.1007/s12567-020-00326-4
7. Douch, K., Wu, H., Schubert, C., et al.: Simulation-based evaluation of a cold atom interferometry gradiometer concept for gravity field recovery. Adv. Space Res. **61**(5), 1307–1323 (2018). https://doi.org/10.1016/j.asr.2017.12.005
8. Elliott, E.R., Krutzik, M.C., Williams, J.R., et al.: NASA's cold atom lab (CAL): system development and ground test status. npj Microgravity **4**, 16 (2018). https://doi.org/10.1038/s41526-018-0049-9
9. European Space Agency.: Steady-state ocean circulation mission. ESA SP **1233**(1), 217 (1999)
10. Frappart, F., Ramillien, G.: Monitoring groundwater storage changes using the gravity recovery and climate experiment (GRACE) satellite mission: a review. Remote Sens. **10**(6), 829 (2018). https://doi.org/10.3390/rs10060829
11. Förste, C., Bruinsma, S. L., Abrikosov, O., et al.: EIGEN-6C4: the latest combined global gravity field model including GOCE data up to degree and order 2190 of GFZ potsdam and GRGS toulouse. GFZ Data Serv. (2014). https://doi.org/10.5880/ICGEM.2015.1
12. Gruber, T., Beutler, G., Fecher, T., et al.: Earth gravity field determination from space—A computational challenge. In: LRZ Linux-Cluster Workshop (2008)
13. Haagmans, R., Siemes, C., Massotti, L., et al.: ESA's next-generation gravity mission concepts. Rend. Lincei Sci. Fis. Nat. **31**, 15–25 (2020). https://doi.org/10.1007/s12210-020-00875-0
14. Ilk, K.H., Flury, J., Rummel, R., et al.: Mass transport and mass distribution in the Earth system. In: Contribution of the New Generation of Satellite Gravity and Altimetry Missions to Geosciences, 2nd edn, GOCE Projektbüro TU München, GFZ Potsdam (2005)
15. Kornfeld, R.P., Arnold, B.W., Gross, M.A., et al.: GRACE-FO: the gravity recovery and climate experiment follow-on mission. J. Spacecr. Rockets **56**(3), 931–951 (2019). https://doi.org/10.2514/1.A34326
16. Mayer-Gürr T., Zehentner N., Klinger B., Kvas A.: ITSG-Grace2014: a new GRACE gravity field release computed in Graz. In: GRACE Science Team Meeting (GSTM), Potsdam (2014)
17. Mayer-Gürr, T., Behzadpour, S., Ellmer, M., et al.: ITSG-Grace2016: monthly and daily gravity field solutions from GRACE. GFZ. (2016). https://doi.org/10.5880/icgem.2016.007
18. Migliaccio, F., Reguzzoni, M., Sansò, F.: Space-wise approach to satellite gravity field determination in the presence of coloured noise. J. Geod. **78**, 304–313 (2004). https://doi.org/10.1007/s00190-004-0396-z
19. Migliaccio, F., Reguzzoni, M., Sansò, F., and Zatelli, P.: GOCE: Dealing with large attitude variations in the conceptual structure of the space-wise approach. In: Proceedings of the 2nd International GOCE user Workshop. Frascati, Rome, Italy, ESA SP-569 (2004)
20. Migliaccio, F., Reguzzoni, M., Sansò, F., Tselfes, N.. On the use of gridded data to estimate potential coefficients. In: Proceedings of the 3rd International GOCE user Workshop, pp. 311–318. Frascati, Rome, Italy, ESA SP-627 (2007)
21. Migliaccio, F., Reguzzoni, M., Sansò, F., Tselfes, N.: An error model for the GOCE space-wise solution by monte carlo methods. In: Sideris, M.G. (eds) Observing our Changing Earth. International Association of Geodesy Symposium, p. 133. Springer, Berlin, Heidelberg (2009). https://doi.org/10.1007/978-3-540-85426-5_40
22. Migliaccio, F., Reguzzoni, M., Batsukh, K., et al.: MOCASS: a satellite mission concept using cold atom interferometry for measuring the earth gravity field. Surv. Geophys. **40**, 1029–1053 (2019). https://doi.org/10.1007/s10712-019-09566-4

23. Mottini, S., Anselmi, A. (2019). Cold atom inertial sensor: mission applications (CAI). Summary report. Thales Alenia Space
24. Pail, R., Bruinsma, S., Migliaccio, F., et al.: First GOCE gravity field models derived by three different approaches. J. Geod. **85**(11), 819–843 (2011). https://doi.org/10.1007/s00190-011-0467-x
25. Pail, R., Bingham, R., Braitenberg, C., et al.: Science and user needs for observing global mass transport to understand global change and to benefit society. Surv. Geophys. **36**, 743–772 (2015). https://doi.org/10.1007/s10712-015-9348-9
26. Papoulis, A.: Probability, Random Variables and Stochastic Processes. McGraw-Hill (1984)
27. Pivetta, T., Braitenberg, C., Barbolla, D.F.: Geophysical challenges for future satellite gravity missions: assessing the impact of MOCASS mission. Pure Appl. Geophys. **178**, 2223–2240 (2021). https://doi.org/10.1007/s00024-021-02774-3
28. Reigber, C., Schwintzer, P., Luhr, H.: The CHAMP geopotential mission. Boll. di Geofis. Teor. ed Appl. **40**(3–4), 285–289 (1999)
29. Reguzzoni, M., Migliaccio, F., Batsukh, K.: Gravity field recovery and error analysis for the MOCASS mission proposal based on cold atom interferometry. Pure Appl. Geophys. **178**, 2201–2222 (2021). https://doi.org/10.1007/s00024-021-02756-5
30. Sorrentino, F., Bodart, Q., Cacciapuoti, L., et al.: Sensitivity limits of a Raman atom interferometer as a gravity gradiometer. Phys. Rev. A. **89**, 023607 (2014)
31. Tapley, B.D., Bettadpur, S., Watkins, M., Reigber, C.: The gravity recovery and climate experiment: Mission overview and early results. Geophys. Res. Lett. **31**(9) (2004). https://doi.org/10.1029/2004GL019920
32. Tino G.M., Kasevich M.A.: Atom interferometry. In: Proceedings of the International School of Physics "Enrico Fermi". vol. 188. ISBN: 978-1-61499-447-3 (print), 978-1-61499-448-0 (online) (2014)
33. Trimeche, A., Battelier, B., Becker, D., et al.: Concept study and preliminary design of a cold atom interferometer for space gravity gradiometry. Class. Q. Gravity **36**(21), 215004 (2019). https://doi.org/10.1088/1361-6382/ab4548

Autonomous Wireless Sensors via Graded Elastic Metamaterials

Jacopo Maria De Ponti

Abstract Amongst the 17th Sustainable and Development Goals (SDGs), it's crucial to ensure access to sustainable and modern energy, as emphasized by the Goal 7. This is not only relevant for large utilities, but also for tiny devices such as wireless sensors that can ubiquitously found in our information driven society. Recent advances in low-power consumption circuitry have enabled ultrasmall power integrated circuits, which can run with extremely low amount of power. For these reasons, energy harvesting can be used to self-power small electronic devices, using ambient waste energy from vibrations. Recent metamaterial technologies allow to dramatically increase the energy available for harvesting, and the operational bandwidth. A large-scale application of metamaterial-based energy harvesting could increase the sustainability in the global energy mix as well as provide improvement in energy efficiency.

Graphical Abstract

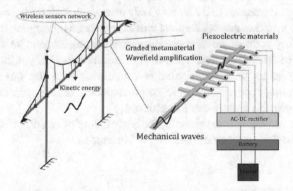

Keywords Affordable and clean energy · Metamaterials · Energy harvesting · Piezoelectricity · Elastic waves · Graded resonators · Sensors

J. M. De Ponti (✉)
Department of Civil and Environmental Engineering, Politecnico di Milano, Piazza Leonardo da Vinci 32, 20133 Milano, Italy
e-mail: jacopomaria.deponti@polimi.it

1 Preliminary Comments and Outlines

Goal 7 of the United Nations SDGs aims at ensuring access to affordable, reliable, sustainable and modern energy. The targets also emphasize the importance of using renewable energy in the global energy mix as well as the improvement in energy efficiency. The possibility to use new forms of clean energy is not only relevant for large utilities, but also tiny devices such as wireless sensors, which generate the data and allow further functionality from self-monitoring and self-configuration to condition monitoring of complex processes. In recent years the world is facing an extraordinary diffusion of the Internet of Things (IOT) concept which is the idea of building smart and autonomous sensors networks which can help us in sensing, understanding, and controlling our environment. For this idea to be effective, new sensors should be small, barely costless, and autonomous. The reduced power requirements of recent small electronic components, makes on-chip energy harvesting solutions a promising alternative to batteries or complex wiring. Amongst others, vibration-based energy harvesting solutions are particularly attractive due to the numerous and continuous sources of vibration present in the environment. However, due to the low amount of energy involved in common ambient vibrations, it is interesting to focus, or trap, waves from a larger region outside the device into a confined region in the near vicinity of the sensor. For these reasons, in order to fully take advantage of this form of energy, it is required a device that: (i) *focus or confine waves*: it is possible to increase the absorbed energy since it comes from a larger spatial region, or due to confinement in specific positions; (ii) *work in a broadband regime*: the energy of common ambient spectra can be completely used, and the performance is less affected by input changing; (iii) *can be easily manufactured*: mass scale production is possible with affordable costs. Several works have been reported in the literature to partially or totally address the aforementioned key requirements. Most of them rely on the design of structuring materials and metamaterials, i.e. engineered systems able to show efficient wave manipulation properties; once the wave is localised, by using electromagnetic, electrostatic, or piezoelectric effects, efficient conversion from elastic to electric energy can be achieved.

To introduce the reader into this field, the basic theory of elastic wave propagation, metamaterials and piezoelectric materials is introduced.

1.1 Mechanical Waves

A *wave* can be defined as the propagation of a disturbance with oscillations about a stable equilibrium configuration. As the disturbance propagates, it carries along amounts of energy that can be transmitted over considerable distances. A *mechanical wave* is a local strain that propagates in a deformable body from particle to particle, by creating local stresses. To create a mechanical wave, two opposed forces that simultaneously counteract and restore equilibrium are required. This is done by the *inertia* and *elastic* forces which correspond, energetically speaking, to the *kinetic* and *potential* energies. If we lose the elastic force, it is not a wave, but motion of mass. If the wave is defined by nodes and anti-nodes, i.e. with fixed peaks amplitude positions in space, it is called *stationary*; contrary, it is called *travelling* or *propagating*. A particular wave is the *plane wave*, that is a wave with constant amplitude for any plane perpendicular to its direction of propagation. It is worth to notice that the direction of propagation does not necessarily coincide with the direction of oscillation of the particles: if equal, the wave is called *longitudinal*, otherwise *transverse*.

An elastic, homogeneous and isotropic continuum, is defined by three parameters: density ρ, Young (or Elastic) modulus E, and Poisson ratio ν. It is reasonable to assume that, in this medium, the wave velocity is $c = c(\rho, E, \nu)$. By adopting the form $c = \xi(\nu)\rho^\alpha E^\beta$, due to dimensional considerations we get: $c = \xi(\nu)\sqrt{E/\rho}$, which means that increasing the elastic parameter or decreasing the inertial one, the wave velocity increases.

The waves supported by an elastic medium, can be identified through the *dispersion relation*, which is a relation between the angular frequency and the wavenumber. If this relation is linear, the medium is called *non-dispersive*, contrary it is called *dispersive*. An important quantity involved in the dispersion relation is the wave velocity, which can be distinguished into *phase velocity* and *group velocity*. The phase velocity is the rate at which the phase of the wave propagates in space, and it is defined as $c_{ph} = \omega/\kappa$. The group velocity is the velocity with which the overall envelope shape of the wave's amplitudes propagates through space (wave packet velocity); it is defined as $c_g = \partial\omega/\partial\kappa$, and it is just the same as the velocity of energy transport of a monochromatic wave [1].

1.2 Metamaterials

The word *metamaterial* etymologically means, from the greek prefix $\mu\epsilon\tau\alpha$, a material with properties beyond what we expect to find in naturally occurring, or conventional materials. Firstly, it is important to notice that there is a strong dependance on the level at which the phenomenon is observed. In wave propagation phenomena (the ones considered here), it is reasonable to take as a reference scale of observation the *wavelength* λ, i.e. the wave spatial period. In other terms, we can adopt a material Representative Volume Element (RVE) of the size of λ for the *homogenised* material, i.e. a homogenous material with equivalent global properties. If λ is much larger than the scale of variation of the internal microstructure (called *unit cell*, in analogy with atoms at smaller scale) we can consider the RVE as homogeneous, as typically done in continuum mechanics. If this RVE shows unusual physical properties (with respect to conventional materials), we denote it as a metamaterial.

Using this convention, a material is considered as a metamaterial if shows unusual properties at strong subwavelength scale. On the contrary, it is simply an inhomogeneous medium. In the setting of elasticity, inhomogeneous media with a periodic structure are usually called Phononic Crystals (PnC), coming from the term phonon, i.e. the quantum vibration, and crystal which suggest the idea of something regularly repeated in space. Specifically, the term phononic is used to say that the phenomenon involves phonons, i.e. vibrations, and that it occurs at the wavelength scale of the order of the unit cell size. The simplest example of a phononic crystal is a spring mass chain [2, 3].

Since the mechanical concept of continuum is meaningful if the behaviour is strongly subwavelength, metamaterials can be based in essence only on resonance effects, in accord with [4]. This is coherent with the seminal work of the group of Ping Sheng at Hkust [5], that provided the first numerical and experimental evidence of a localised resonant structure for elastic waves propagating in three-dimensional arrays of thin coated spheres. Adopting this interpretation, metamaterials, contrary to phononic crystals, can be even aperiodic. However, they are usually periodically defined, due to the peculiar properties given by periodicity, as well as reduced computational complexity and the existence of analytical closed form solutions. The work of Liu in acoustics opens the door to the design of elastic metamaterials, but this concept was preceded by important discoveries in electromagnetism and optics.

In 1967, the Russian physicist Victor Veselago published a visionary paper [6] in which electromagnetic media with simultaneously negative permittivity ϵ and magnetic permeability μ were shown to be characterized by a negative refractive index of refraction. He showed that a slab of a negative refractive index material can act as a flat convergent lens that images a source on one side to a point on the other. This discovery remained an academic curiosity for almost three decades, until the British physicist. John Pendry [7, 8] proposed effective designs of structuring materials with negative ϵ and μ, and experimental demonstrations were done at the GHz frequencies by a handful of photonic groups in the United States (2000) [9]. As previously emphasised, these materials are structuring at subwavelength scale (typically $\lambda = 10$), hence it is possible to regard them as nearly homogeneous. The term metamaterial, coined by Walser [10], describes such periodic structures when one can average their properties, which are strongly dispersive and anisotropic [4].

Metamaterials, and in general structuring materials, due to their unique properties to guide the propagation of elastic waves and focus their energy, can be adopted to enhance vibration-based energy harvesting [11]. These features have also been exploited for the design of innovative actuators and sensors, and elements of logic circuitry based on the propagation of elastic waves [12, 13].

1.3 Piezoelectricity

Piezoelectricity is the ability of a material to develop an electric charge in response to an applied mechanical stress (direct piezoelectric effect) and vice-versa (inverse piezoelectric effect). The term comes from the Greek words $\pi\iota\epsilon\zeta'\epsilon\iota\nu$, which means to squeeze or press and $\eta'\lambda\epsilon\kappa\tau\rho\nu$, meaning amber, an ancient source of electric charge. It was firstly discovered by the French physicists Jacques and Pierre Curie in 1880, which demonstrated the direct piezoelectric effect in quartz and in other crystalline materials in the natural state. The piezoelectric effect is due to the peculiar crystalline structure of such materials, with no inversion symmetry. Electrical dipoles within the piezoelectric material are responsible for the creation of a potential difference across the material, when the top and bottom layers are connected to electrodes. When the material is in the unstressed state, it is neutrally charged, since the positive and negative charges balance each other. Contrary, the application of a stress, changes the position of the charges, thus modifying the dipole moment.

The most common piezoelectric materials are ceramics, and specifically Aluminium. Nitride (AIN) and Lead Zirconate Titanate (PZT) due to their piezoelectric and manufacturability qualities. In order to describe the behaviour of piezoelectric materials in the setting of continuum mechanics, the electromechanical coupling enters in the constitutive laws [14]:

$$\begin{cases} T_{ij} = c^E_{ijkl} S_{kl} - e_{kij} E_k \\ D_i = e_{ikl} S_{kl} + \varepsilon^S_{ik} E_k \end{cases} \tag{1}$$

where c^E_{ijkl} is a 4th order elastic stiffness symmetric tensor evaluated at constant electric field; e_{kij} is a 3rd order tensor of the piezoelectric stress constants, and ε^S_{ik} is a 2nd order tensor of the dielectric constants at constant strain. Equation (1) is the *e-form* of the piezoelectric constitutive equations, in which the strain and the electric field are used as coupling variables. Alternatively, it is possible to use the stress instead of the strain, obtaining the *d-form* of the piezoelectric constitutive equations:

$$\begin{cases} S_{ij} = s^E_{ijkl} T_{kl} + d_{kij} E_k \\ D_i = d_{ikl} T_{kl} + \varepsilon^T_{ik} E_k \end{cases} \tag{2}$$

where s^E_{ijkl}, d_{kij} and ε^T_{ik} are the 4th order elastic symmetric compliance tensor at constant electric field, the 3rd order tensor of the piezoelectric strain constants and the 2nd order tensor of the dielectric constants at constant stress respectively. The peculiarity of piezoelectricity is given by 3rd order tensors e_{kij} and d_{kij}, that couple mechanical and electrical quantities.

2 Graded Elastic Metamaterials

A crucial aspect to manipulate waves is the possibility to create *band gaps*, i.e. frequency ranges for which the wave propagation is forbidden. From a different perspective, in those frequency ranges, only *evanescent waves* exist. This means that the wave does not propagate in the medium, and the energy is spatially concentrated in the vicinity of the source. A hallmark of an evanescent wave is that there is no energy flow in the considered region, with a Poynting vector (i.e. the directional energy flux) equal to zero. By operating within or close to a band gap, waves can be guided and focused, and this can be useful to locally enhance the harvestable energy.

Several approaches have been reported in the literature to focus energy using structuring materials, dynamic induced anisotropy or metamaterials, usually combined with smart materials for energy harvesting and sensing. Most of them rely on the creation of mirrors and funnels [11], defects in phononic crystals [15, 16], local resonators [17–19] or lenses [20, 21].

The intensive research on this topic reveals its relevance inside the scientific community, motivating novel approaches to overcome current missing or limits of the existing solutions. While parabolic mirrors, funnels and lenses are able to focus elastic energy from large regions of space, one of the main drawbacks is the capability to use them for finite structures, where waves are intrinsically confined, and boundary reflections can partially or totally alter the wavefield. On the other hand, broadband behaviour is difficult to be achieved; a change in frequency reflects significantly on the wavelength, resulting in ineffective interactions between the wave and the device. On the contrary, devices based on local resonance are less affected by boundary effects and allow for an interaction with large wavelengths. While this is a strong advantage for real applications, the periodic repetition of identical resonators reduces the broadband capability of these systems. Moreover, close to a local resonance bandgap, only the resonators near the input region are expected to store most of the energy, reducing the overall efficiency of the device. For these reasons, the usage of different types of resonators to control and confine waves looks promising, due to their capabilities to take energy from broadband signals at strong subwavelength scale. Moreover, by gradually varying the medium effective properties, it is possible to modify the waves, that are spatially compressed and amassed, with a strong amplitude enhancement.

The advantages of such designs, based on the so-called *rainbow effect* [22–27] have been demonstrated in different contexts, from electromagnetism to acoustics and elasticity, but not extensively for vibration-based energy harvesting. Theoretical models, together with numerical finite elements analyses and experiments, demonstrate that graded metamaterials are excellent candidates for vibration-based energy harvesting. These findings, numerically demonstrated and experimentally validated at meso scale, can be even extended to micro scale, for the implementation of next generation vibration energy harvesting devices.

2.1 Numerical Analyses on Slow Flexural Waves

A graded array is formed by smoothly varying a particular parameter or set of parameters of neighboring elements in space through a specific design of consecutive unit cells. Figure 1a shows an elastic beam with attached an array of cantilevers of linearly increasing lengths.

The beam and the resonators are made of aluminium with Young modulus $E_a = 70$GPa, Poisson ratio $\nu_a = 0.33$ and density $\rho_a = 2710$kg/m^3. The beam is 500mm

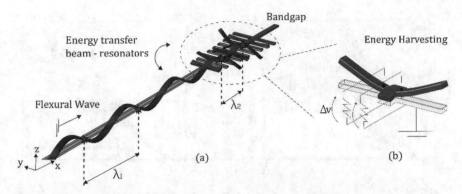

Fig. 1 Schematic of the graded linear array of resonators for energy harvesting. **a** By exciting the waveguide with a flexural wave, energy is efficiently transferred to the array of resonators. Such interaction reduces both the amplitude and the wavelength of the waves along the waveguide within the array. Leveraging on this energy transfer mechanism, the elastic energy in the resonators can be used for piezoelectric energy harvesting (**b**)

long, 7mm wide and 2mm thick. The array is made of 9-unit cells of size $a = 15$mm, with a linear grading law for the lengths of the resonators, from 16.75 to 27.75mm, resulting in a grading angle of approximately $5.2°$.

By spatially varying the resonance frequency of the resonators attached to the beam [28–32] waves slow down with a reduction of both amplitude and wavelength. Differently with respect to the acoustic wave compression [24], the array of resonators progressively absorbs energy from the beam, allowing for a wave amplitude reduction in the beam inside the array [29].

The concurrent amplitude and wavelength reduction is a hallmark of energy transfer between the main structure and the resonators and can be used for energy harvesting purposes. Figure 1b shows the arrangement of piezoelectric patches and the electric circuit employed to transduce electric due to resonators motion in a tailored position along the beam. This is obtained exploiting the 31-mode of the piezoelectric patches connected to a resistive load, to effectively harvest the elastic energy stored inside the target resonators.

The wave propagation properties of the system can be rigorously inferred by looking at the dispersion curves of a given cell inside the array. Provided the grading is gentle enough and provided the number of unit cells is sufficient, the global behaviour of the whole array can be deduced from the local dispersion curves of the constituent elements [26]; in this way, the desired spatial selection by frequency, i.e. the rainbow behaviour of the system, is determined from the locally periodic structure at a given position. Figure 2a shows the numerical dispersion curves for the cell number 7 (where the cell numbering in the array goes from 1 for the shortest to 9 for the longest). These dispersion curves are computed along the 1D irreducible Brillouin Zone using the finite elements software Abaqus®, that incorporates the Bloch phase shift via Bloch-Floquet periodic boundary conditions. The spatial properties of the wavefield can be deduced from the local dispersion curves at a given frequency,

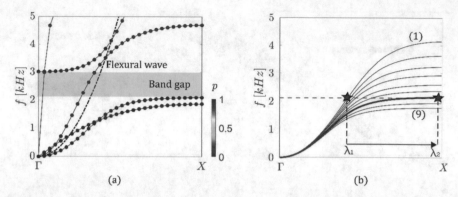

Fig. 2 **a** Numerical dispersion curves for a periodic array of identical resonators of fixed length (in this case the resonator length is 25 **mm**); scatter points colors represent the wave polarization (red corresponding to the vertical motion, i.e. bending of the resonator). **b** In-phase bending mode for different resonators inside the array. Moving from short (1) to long (9) resonators at a given frequency the group velocity and the wavelength decrease, until the bandgap opening

as shown in Fig. 2b. By increasing the length of the resonators along the spatial dimension, i.e. moving from (1) to (9), the dispersion curves shift towards lower frequencies. As a result, by fixing the frequency, the group velocity, $c_g = \partial\omega/\partial\kappa$, smoothly reduces until zero. Such effect allows to slow down elastic waves inside the array and to confine waves in different positions depending on frequency. In addition, since the zero-group velocity mode occurs at the band edge, it can couple with a backward propagating mode, which is typical of rainbow reflection [30].

2.2 Experiments on Graded Waveguides for Energy Harvesting

A peculiar property of this system is the capability to slow down array guided waves as they transverse the array. Such phenomenon allows for a longer interaction between the wave and the resonators, locally increasing the amplitude of the wavefield inside the resonators [32]. To validate this effect, and the implications in terms of energy harvesting, experimental tests are performed in narrow and broad-band frequency regime. Figure 3 shows the experimental setup used for testing.

At the right boundary, a LDS v406 electrodynamic shaker is rigidly connected to the beam through a thick aluminium plate with high strength adhesive, to provide excitation. At the opposite boundary, the structure is suspended through elastic cables that do not affect the dynamics of the system. The wavefield on the elastic beam is measured through a Polytec 3D Scanner Laser Doppler Vibrometer (SLDV), which is able to separate the out-of-plane velocity field in both space and time. The input is synchronously start with the acquisition which, in turn, is averaged in time to decrease the noise. We experimentally show the rainbow effect in the linear array by

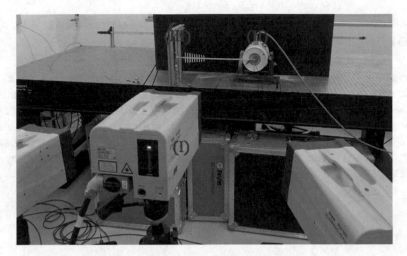

Fig. 3 Experimental setup. The wavefield is measured using a 3D laser vibrometer (I). The waveguide is excited using an electrodynamic shaker (II), while the opposite side is suspended through elastic cables (III)

applying a broadband frequency sweep in the range 1.6–4.2 kHz. Figure 4a shows the space-frequency analysis of the experimental data. Depending on the frequency, waves stop at different spatial positions, corresponding to the band gap opening.

Moreover, we notice that the amplitude and the wavelength of the mode shapes decrease inside the array, until the amplitude vanishes in correspondence of the position of the resonating element, which is well predicted by numerical results (dashed white line). We then quantify the advantages of such mechanism for energy

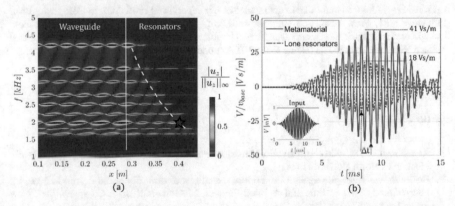

Fig. 4 Space-frequency analysis of the experimental data superimposed to numerical predictions (dashed white line) from dispersion curves. **b** Open circuit output voltage for the graded array and single cell configuration (i.e. the same harvester but without the graded metamaterial), corresponding to the input excitation frequency marked with the red star in (**a**). The graded metamaterial amplifies the output voltage, i.e. the energy harvesting capabilities of the device

harvesting by placing piezoelectic PZT-5H patches ($E_p = 61$ GPa, $\nu_p = 0.31$, $\rho_p = 7800$ kg/m^3, dielectric constant $\frac{\varepsilon_{33}^T}{\varepsilon_0} = 3500$, and piezoelectric coefficient $e_{31} = -9.2$ C/m^2) at the position of the 7th cell, denoted with the red star in Fig. 4a.

Figure 4b shows the mean output open circuit voltage for the single cell (i.e. the same resonators but without the graded metamaterial), and graded array, normalized by the measured input velocity, to make sure that the results are displayed under the same conditions. We observe that the graded array gives a mean normalized peak voltage of 41 Vs/m which is 56% higher than the single cell. We notice that such peak is reached with a delay Δt of approximately 1.3 ms, which is justified by the smooth reduction of the group velocity inside the array.

3 Conclusions

In conclusion, potential advantages in using graded metamaterials for efficient elastic energy confinement have been demonstrated both numerically and experimentally. The graded metamaterial capability of slowing down waves enables a strong energy transfer to the resonators, which then reflects in enhanced energy harvesting performances. This mechanism can be adopted to create more efficient energy harvesting systems able to self-power, or at least compensate, the power consumption of small electronic devices, using ambient waste energy from vibrations. The possibility to substitute batteries and their chemical waste, with fully sustainable vibration-based energy harvesting systems can play an important role toward Goal 7 of the SDGs for the next generation of affordable and clean energy sensors and small electronic devices.

Acknowledgements I would like to acknowledge my supervisors Prof. Alberto Corigliano, Prof. Raffaele Ardito, Prof. Francesco Braghin and Prof. Richard Vaughan Craster for their help throughout this thesis. This research was funded by the H2020 FET-proactive project Metamaterial Enabled Vibration Energy Harvesting (MetaVEH), under Grant Agreement No. 952039.

References

1. Achenbach, J.D.: Wave propagation in elastic solids. North-Holland Ser. Appl. Math. Mechanics North Holland (1984)
2. Brillouin, L.: Wave propagation in periodic structures: electric filters and crystal lattices. In: International Series in Pure and Applied Physics, McGraw-Hill Book Company, Inc., New York (1946)
3. Kittel, C.: Introduction to Solid State Physics, 8th edn. Wiley, New York (2004)
4. Craster, R.V., Guenneau, S.: Acoustic Metamaterials: Negative Refraction, Imaging, Lensing and Cloaking. Springer Series in Materials Science, Berlin (2013)
5. Liu, Z., Zhang, X., Mao, Y., Zhu, Y.Y., Yang, Z., Chan, C.T., Sheng, P.: Locally resonant sonic materials. Science **289** (2000)

6. Veselago, V.G.: The electrodynamics of substances with simultaneously negative values. Soviet Phys. Uspekhi 10 (1968)
7. Pendry, J.B., Holden, A.J., Stewart, W.J., Youngs, I.: Extremely low frequency plasmons in metallic mesostructures. Phys. Rev. Lett. 76 (1996)
8. Pendry, J.B., Holden, A.J., Robbins, D.J., Stewart, W.J.: Magnetism from conductors and enhanced nonlinear phenomena. IEEE Trans. Microw. Theory Tech. 47 (1999)
9. Smith, D.R., Willie, J., Padilla, D.C., Vier, S.C., Nemat-Nasser, Schultz, S.: Composite medium with simultaneously negative permeability and permittivity. Phys. Rev. Lett. 84 (2000)
10. Walser, R.M.: Metamaterials: what are they? What are they good for? Bull. Am. Phys. Soc. (2000)
11. Carrara, M., Cacan, M.R., Toussaint, J., Leamy, M.J., Ruzzene, M., Erturk, A.: Metamaterial-inspired structures and concepts for elastoacoustic wave energy harvesting. Smart Mater. Struct. 22 (2013)
12. Torrent, D., Sánchez-Dehesa, J.: Acoustic metamaterials for new two-dimensional sonic devices. New J. Phys. 9 (2007)
13. Bilal, O.R., Foehr, A., Daraio, C.: Bistable metamaterial for switching and cascading elastic vibrations. In: Proceedings of the National Academy of Sciences of the United States of America, vol. 114 (2017)
14. Erturk, A., Inman, D.J.: Piezoelectric Energy Harvesting. Wiley, West Sussex (2011)
15. Wu, L.Y., Chen, L.W., Liu, C.M.: Acoustic energy harvesting using resonant cavity of a sonic crystal. Appl. Phys. Lett. 95 (2009)
16. Qi, S., Oudich, Y., Li, M., Assouar, M.: Acoustic energy harvesting based on a planar acoustic metamaterial. Appl. Phys. Lett. 108 (2016)
17. Gonella, S., To, A.C., Liu, W.K.: Interplay between phononic bandgaps and piezoelectric microstructures for energy harvesting. J. Mech. Phys. Solids 57 (2009)
18. Ahmed, R.U., Banerjee, S.: Low frequency energy scavenging using sub-wavelength scale acousto-elastic metamaterial. AIP Adv. 4 (2014)
19. Sugino, C., Erturk, A.: Analysis of multifunctional piezoelectric metastructures for low-frequency bandgap formation and energy harvesting. J. Phys. D: Appl. Phys. 51 (2018)
20. Tol, S., Degertekin, F.L., Erturk, A.: Gradient-index phononic crystal lens-based enhancement of elastic wave energy harvesting. Appl. Phys. Lett. 109 (2016)
21. Zareei, A., Darabi, A., Leamy, M.J., Alam, M.R.: Continuous profile flexural GRIN lens: focusing and harvesting flexural waves. Appl. Phys. Lett. 112 (2018)
22. Tsakmakidis, K.L., Boardman, A.D., Hess, O.: 'Trapped rainbow' storage of light in metamaterials. Nature 450 (2007)
23. Zhu, J., Chen, Y., Zhu, X., Garcia-Vidal, F.J., Yin, X., Zhang, W., Zhang, X.: Acoustic rainbow trapping. Sci. Rep. 3 (2013)
24. Chen, Y., Liu, H., Reilly, M., Bae, H., Yu, M.: Enhanced acoustic sensing through wave compression and pressure amplification in anisotropic metamaterials. Nat. Commun. 5 (2014)
25. Romero-García, V., Picó, R., Cebrecos, A., Sánchez-Morcillo, V.J., Staliunas, K.: Enhancement of sound in chirped sonic crystals. Appl. Phys. Lett. 102 (2013)
26. Colombi, A., Colquitt, D., Roux, P., Guenneau, S., Craster, R.V.: A seismic metamaterial: the resonant metawedge. Sci. Rep. 6 (2016)
27. Colombi, A., Ageeva, V., Smith, R.J., Clare, A., Patel, R., Clark, M., Colquitt, D., Roux, P., Guenneau, S., Craster, R.V.: Enhanced sensing and conversion of ultrasonic Rayleigh waves by elastic metasurfaces. Sci. Rep. 7 (2017)
28. De Ponti, J.M., Colombi, A., Ardito, R., Braghin, F., Corigliano, A., Craster, R.V.: Graded elastic metasurface for enhanced energy harvesting. New J. Phys. 22, 013013 (2020)
29. De Ponti, J.M., Colombi, A., Riva, E., Ardito, R., Braghin, F., Corigliano, A., Craster, R.V.: Experimental investigation of amplification, via a mechanical delay-line, in a rainbow-based metamaterial for energy harvesting. Appl. Phys. Lett. 117, 143902 (2020)
30. Chaplain, G.J., Pajer, D., De Ponti, J.M., Craster, R.V.: Delineating rainbow reflection and trapping with applications for energy harvesting. New J. Phys. 22, 063024 (2020)

31. Chaplain, G.J., De Ponti, J.M., Aguzzi, G., Colombi, A., Craster, R.V.: Topological rainbow trapping for elastic energy harvesting in graded Su-Schrieffer-Heeger systems. Phys. Rev. Appl. **14**, 054035 (2020)
32. De Ponti, J.M.: Graded Elastic Metamaterials for Energy Harvesting. Springer International Publishing, Switzerland (2021)

Limit Analysis for Masonry Vaults and Domes of Cultural Heritage

Maria Chiara Giangregorio

Abstract The scope of this dissertation is the study of masonry shell structures of cultural heritage through limit analysis. The final aim is to set up computational tools, integrated with experimental observations, for use in the interpretation of observed crack patterns and to study the safety of these structures. The kinematic approach is developed to calculate the limit load and the three-dimensional collapse mechanism of symmetric and skew barrel vaults. The effect of structural strengthening is examined as well. A static discrete model is developed for double curvature and polygonal domes, applying equilibrium at the blocks and strength criteria at interfaces. This allows to study a cracked masonry dome, evaluating the collapse multiplier and mechanism by considering the shell flexural response mechanism. This is used also to analyse the actual behaviour in membrane regime of a dome on reaching the maximum load before cracking. Finally, the polygonal dome supporting a tower of the Chiaravalle Abbey in Milan is analysed.

Graphical Abstract

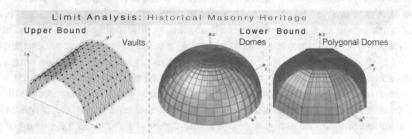

Keywords Cultural heritage · Limit analysis · Upper bound · Lower bound · Vault · Dome · SDGs 11

M. C. Giangregorio (✉)
Department of Civil and Environmental Engineering, Politecnico di Milano, Piazza Leonardo da Vinci 32, 20133 Milano, Italy
e-mail: mariachiara.giangregorio@polimi.it

M. Antonelli and G. Della Vecchia (eds.), *Civil and Environmental Engineering for the Sustainable Development Goals*, PoliMI SpringerBriefs,
https://doi.org/10.1007/978-3-030-99593-5_6

1 Protect and Safeguard the word's Cultural Masonry Heritage

1.1 Introduction

Among the 17 Sustainable Development Goals (SDGs) this dissertation concerns the 11th goal: 'Make cities inclusive, safe, resilient and sustainable'. The focus is on the target 11.4 'Strengthen efforts to protect and safeguard the world's cultural and natural heritage'. Here the method of limit analysis is used for the study of historical masonry structures concerning cultural heritage such as domes and vaults. In order to do that, the first aim of the work is to develop simple discrete algorithms based on the two theorems of limit analysis (static and kinematic theorem). The models introduced are here shown and used for the structural study of barrel vaults, rotational domes, and polygonal domes. These algorithms use optimization of a load multiplier, to calculate the collapse multiplier and the collapse mechanism, but also the multiplier leading to crack initiation. The study of the actual state of the structure with the static method, that allows to find an equilibrated solution based on membrane behaviour, is one of the aims of the work. This is possible by considering the tensile strength at the block interfaces. To consider the three different states of the structure (un-cracked, cracked and collapse) allows a better understanding of the safety state of the structure. It must be underlined that the collapse of cultural heritage structures is strictly related to the cracked state. The focus on the geometry and its proportion reported in the work aims at understanding the role they play in the structural behaviour. Thanks to the simplicity of these models, it was possible to compare different geometries, load conditions and strength criteria, without any relevant computational effort. Two interesting typology of dome was investigated: hemispherical and polygonal domes. This allows to analyse the different behaviour of single and double curvature shells. This work is part of a research project that aims at developing an interdisciplinary methodology for structural analysis of historical masonry elements, to understand the causes of existing damage, verify the structural safety and study possible structural strengthening. This is pursued by using an integrated method of limit analysis with the survey and measurement of geometry, considering deformations, crack patterns and other damage. The multidisciplinary method includes the study of historical construction phases, the analysis of the actual configuration, with the survey of the building and the annotation of the damage, and the structural analysis, with the suggested method of limit analysis.

1.2 The Geometry Considered

In the ensemble of cultural masonry heritage, different shell shapes give different structural response. Since there is a close correlation between the geometry of a surface and its main failure mechanisms, the focus on geometry is a topic of interest.

A vault can be obtained by translation of the arch, as in barrel vaults (Fig. 1.) or by revolution around a vertical axis, as in the hemispherical dome (Fig. 2). The different typologies of vaults can be subdivided into "simple vaults" and "composite vaults". The "simple vaults" are those where the surface does not have solutions of continuity because there is no point in which the curvature changes as: (i) Barrel vault, (ii) Hemispherical dome, (iii) Sail vault (deriving from the dome). The "composed vaults" are those realized through the intersection of two or more parts of a simple vault, so that the surface has several zones in which the curvature changes. In order to describe a composite vault, it is necessary to observe the elements obtained by sectioning a barrel vault along the two diagonals. There will be two "spindles" and two "nails". If four nails are combined a cross vault is obtained, while with four spindles a cloister vault is obtained. The typical typologies are: (i) Cross vaults/groin vault, (ii) Cloister vaults, (iii) Ribbed vaults (cross vault with ribs) (iv) Polygonal vaults, (v) Stellar vaults, (vi) Fan vaults. In this work the focus is on: (i) barrel vault, (ii) dome, (iii) polygonal dome.

1.3 Barrel Vault

The barrel vault (Fig. 3), is the simplest typology of vault and is the base for many vaults of more complicated shape. It is composed of the translation of an arch in its

Fig. 1 Barrel vault with the geometry of the construction

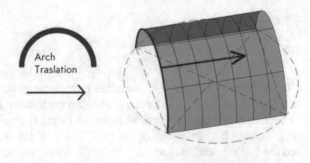

Fig. 2 Hemispherical dome with the geometry of the construction

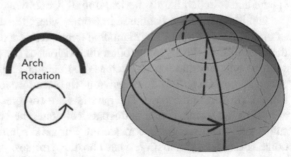

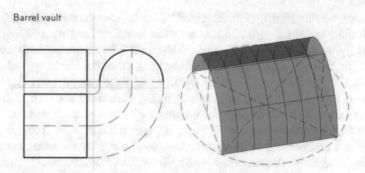

Fig. 3 Barrel vault, with square base and rounded arch

perpendicular direction. It is mostly used as coverage of monumental buildings of a civil and religious nature.

The barrel vault has a semi-cylindrical intrados surface, but the vertical section can be ogival section if it is based on a translation of a pointed arch, but also parabolic or any other curved shape. In plan it is possible to have different quadrilateral shapes, and two opposite parallel walls are the typical vertical supports. A barrel vault subjected to self-weight only has "arch behaviour", but with different load positions or in presence of soil settlements, it could have a different behaviour, such as torsion. If the limit stress is reached, cracks will appear in the direction of the tension, on the intrados or on the extrados, as in the hinged mechanism for the arch.

1.4 Hemispherical Dome

The geometry of a dome is a revolution membrane generated by an arch that rotates around its central axis. It is a convex rounded vault with a horizontal section that may be circular elliptical or polygonal. In vertical section the dome may be hemispherical (Fig. 4), circular, parabolic, or partly elliptical. If the horizontal section is a circle and the vertical section is a round arch, the dome assumes the form of a portion of a sphere and is called hemispherical dome. A dome may be smooth or coffered as in the famous case of the Pantheon in Rome, where the dome is supported upon walls of circular plan. One of the main characteristics of a dome with respect to vaults is that of minor thickness to the other dimensions of the structure. In membrane static solution only internal forces, such as the meridian and parallel, lying on the tangent plane are considered. In this behaviour the compression along the meridians grows from the crown to the base. The parallels are compressed in the key and in tension at near to the springer, in a hemispherical dome the change is at 51°50′° if a dome subjected to self-weight is considered. For axial symmetrical load a hemispherical dome, along the same parallel, has constant stresses.

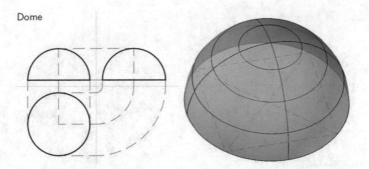

Fig. 4 Hemispherical dome

The problem of the stability of a dome is an ancient topic, due to the tensional state that determines lack of stability over time; when the tensile strength is overcome, cracks will appear at the base of the dome along the meridians. In such cases there is no more membrane behaviour, but if a series of half arches is considered, by dividing the dome into slices cut along the meridian plane, and if it is possible to draw a thrust line inside every arch, the dome can be considered safe [14]. The simplification of the behaviour of a cracked dome by dividing it into a series of arches was firstly developed by Poleni for his study on the Vatican Dome [20], who made one of the first examples of a drawing of crack pattern on a dome was applied to the dome of San Pietro in Rome. Here an algorithm is formulated that allows to analyse domes in their actual state and in their collapse state, considering different configurations. These different configurations concern geometrical parameters such as span, thickness, number of slices, different load conditions, also considering the effect of strengthening.

1.5 Polygonal Dome

The polygonal domes (Fig. 5) are a very particular type of dome, divided into slices. Unlike simple domes, which have a double-curved structure, resembling the shape of a spherical surface, the slices of polygonal domes are single-curved and derive from the section of a cylindrical surface. They are called polygonal domes because if sectioned with a horizontal plane parallel to the floor, the resulting section has the shape of a regular polygon. This structure geometry is similar to a pavilion vault, but with more than four edges. As it was demonstrated in this work, these domes have the characteristic of concentrating the loads in the corners of the polygon, developing concentrated and pushing forces on pointed vertical supports, such as pillars. A support on the structural behaviour is given by the diagonal ribs, often present in the polygonal domes. These typologies of dome are little analysed, Flügge [10] proposed an infinitesimal equilibrium solution for a dome subjected to self-weight only. Here a discrete algorithm is formulated that allows to analyse polygonal domes in their actual

Polygonal dome

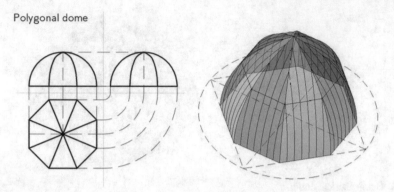

Fig. 5 Polygonal dome

state considering different configurations. These different configurations concern geometrical parameters such as span, thickness, number of slices, different load conditions (Fig. 6).

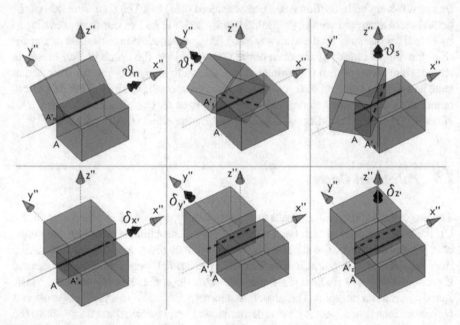

Fig. 6 Rotations and displacements considered in three-dimensional space

2 The Limit Analysis Algorithms for Shell Structures

2.1 The Kinematic Algorithm Applied to Barrel Vaults

With respect to the kinematic theorem of limit analysis applied to shell structures, the aim of this chapter is to study the behaviour of masonry shell structures in their collapse mechanism. The kinematic method of limit analysis is applied to a three-dimensional shell problem with a formulation inspired by 'discontinuity layout optimization' (DLO), that was formulated for plane plasticity problems [22]. The method was successfully used for geotechnical [22], plate and slab problems of arbitrary geometry [11], and on three dimensional problems [13]. Here the innovation is to use it to analyse shell structures, also considering the effect of structural strengthening. An application of the kinematic theorem of limit analysis to barrel vault problems is here addressed. The main results are related to: (i) identification of the collapse mechanism; (ii) identification of the ultimate load carrying capacity: analysis of the amount of live load that can be applied before structural collapse; (iii) analysing strengthened barrel vaults.

Different yield domains and different behaviour of the material are considered. A case with infinite compression strength, no tensile strength and no sliding as stated by Heyman [15]. Then tensile strength different from zero and even the non-symmetric behaviour of a reinforced vaults is contemplated, considering different values of tensile strength according to the different types of reinforcement used. The contribution of the friction angle, mostly in the case of a torsion mechanism is also considered. The plastic dissipation, investigated for reinforced vaults, is governed by three dimensional hinges coupled with normal displacements at the interfaces. Strength criteria formulated in terms of force are opportunely converted into a displacement flow rule to fulfil the formulation that characterizes the DLO method.

This kinematic algorithm, which involves the achievement of collapse, was validated thanks to experimental tests concerning masonry vaults collapse reported in articles that were found in the literature. Some experimental tests on collapse of masonry barrel vaults are considered the benchmark, in particular the barrel rectangular vault tested by Vermeltfoort [27] was considered, that was also used as benchmark for the numerical simulations of [19]. From these experimental tests even barrel vaults with skew arches were investigated, to validate the behaviour of collapse due to torsion. In addition, the experimental tests performed by Girardello [12] allow to validate the work on reinforced vaults, considering two different materials for strengthening.

2.2 The Static Algorithm Applied to Rotational Domes

A discrete model is formulated, based on the static theorem of limit analysis to analyse domes. The structure is divided into rigid blocks, with forces acting on the interfaces. The equilibrium is imposed at each block and a resistance criterion is imposed at each interface. Two different behaviours are considered, membrane behaviour to analyse the actual state and flexural behaviour to analyse the collapse state. The internal forces of redundant structure with crack initiation load multiplier and the collapse load multiplier with corresponding mechanism are obtained after a simplex algorithm imposed. It is possible to analyse the structural behaviour considering different variables, such as geometry, loads variations and the presence of a reinforcement. Two different typologies of shell are analysed: the rotational dome and the polygonal dome. The former has a double curvature and a circular base, while the latter has a single curvature, in zenith direction.

In the literature, regarding the hemispherical domes, the solutions of differential equilibrium equations based on membrane theory have been acquired for self-weight and a few other loading conditions [10]. Membrane theory considers the stresses acting on an infinitesimal element, considering meridian and hoop stresses, and shear forces tangent to the mid surface. In this work it is proven that it is possible to analyse these structures using a discrete model, applying the static limit analysis theorem. The focus here is on the interaction between the geometry of the dome and its structural behaviour. Double curvature domes, rotational domes with round arch and pointed arch shape, are analysed with the static model in their membrane behaviour (Fig. 7),

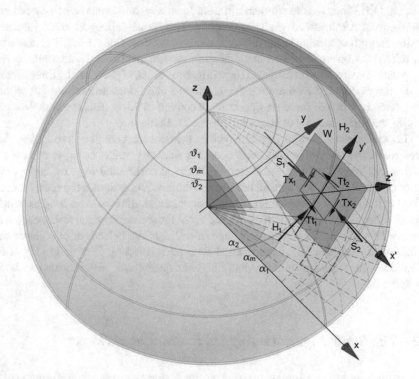

Fig. 7 Scheme of a rotational dome block in membrane behaviour

focusing on the response related to the different geometries. Furthermore, with the same discrete algorithm configuration, and using a flexural approach (Fig. 8), rotational domes are analysed up to collapse, finding the corresponding load multiplier and the collapse mechanism.

2.3 The Static Algorithm Applied to Polygonal Domes

The analysis of polygonal domes is investigated, in their membrane behaviour, developing the static algorithm for polygonal domes that have straight line in the hoop force direction for their single curvature shell. Since the geometry of the surface between

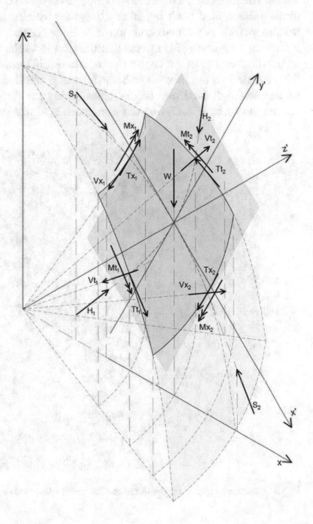

Fig. 8 Forces acting on a block of the double curvature dome, flexural behavior

double curvature and single curvature domes differs, consequently, the response of the structure to load is also different. With the aim of describing the membrane behaviour of polygonal domes some geometric characteristics have been modified in the static model. The focus here is on the interaction between the geometry of the dome and the structural behaviour (Fig. 9).

In the literature, Flügge [10] directly analyses the polygonal domes subjected to self-weight, solving the differential equations of the membrane state for the specific case. D'Ayala [9, 23] and Como [8] consider pavilion vaults, that have a similar structure with single curvature surface, starting from their geometric configuration as portions of cylinders that concentrate most of the part of the loads along the diagonal ribs. In the work of D'Ayala and Tomasoni [9], a theory was formulated, based on lower bound approach of limit analysis, for the structural analysis of pavilion vaults. The formulation includes an optimized solution that allows to find the optimal thrust surface of a vault by minimizing the distance between thrust surface and middle surface. A strength criterion with finite friction is considered among block interfaces considering 3D effects that develop in vaults of complex geometry such as pavilion vaults, in the angular parts where different surfaces are connected. In this dissertation, the membrane behaviour of a single curvature dome is analysed, in a structure divided into blocks, considering the thrust surface coincident with the middle surface, analysing the maximum load that could be reached before the

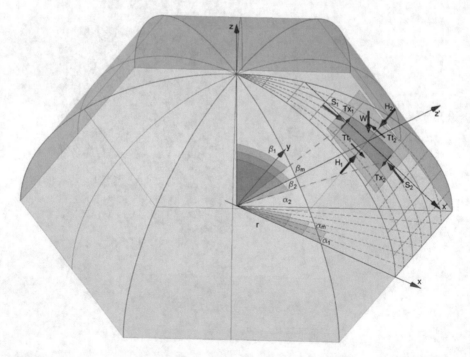

Fig. 9 Reference systems for the dome section and the forces acting on a block of polygonal dome

structure begins to crack. At this point the discrete static algorithm is applicable to polygonal domes: (i) to evaluate the action of a dome on underlying structures for single curvature dome; (ii) to solve the problem of the equilibrium of the diagonal rib; (iii) to compare maximum forces with crack positions both for double curvature dome and for single curvature dome. In the existing case of Chiaravalle abbey in Milan described in [6] the work allowed to: (i) consider load different from the self-weight adding the load of a lantern tower; (ii) to analyse and compare the behaviour of a polygonal dome in membrane behaviour and in a cracked state, by the integration of the structural analysis and the crack pattern.

2.4 Application to Real Case Studies

Some analytical cases are considered in order to analyse the effect of the geometry on the structural response. Amongst the topics dealt with are: (i) different spans; (ii) dome based on a round arch or on a pointed arch; (iii) the difference derived from tapering i.e. a reduction of the section thickness of the dome; (iv) the effects of different load conditions, such as the presence of a lantern; (v) the effect of proper structural strengthening with circular tie rods. Some existing cases of rotational domes with a lantern are analysed: the churches of (i) Sant' Agnese in Agone in Rome [3, 17] (Fig. 10); (ii) San Nicolò l'Arena in Catania [17, 18] (Fig. 11); (iii) San Giorgio in Ragusa [16] (Fig. 12).

Then the polygonal dome of Chiaravalle abbey in Milan is analysed. In this case the weight of a lantern tower called Ciribiciaccola is analysed with surveys, drawings and 3D model, showing the importance of the integrated approach to understand the geometrical configuration. After the analysis of the crack pattern, it is possible to analyse the behaviour of this dome. The analysis of geometry and materials is used as input to be included in the discrete algorithm and the application to Chiaravalle abbey also shows the information deriving from the crack pattern for the structural

Fig. 10 Sant'Agnese in Agone

Fig. 11 San Nicolò L'Arena

Fig. 12 San Giorgio

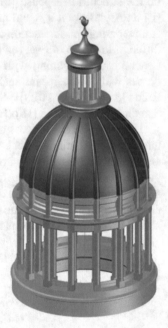

analysis. The analysis of the dome shells both cracked and un-cracked are analysed and compared in their different structural response. Finally, the equilibrium of the rib with radial wall is calculated to evaluate their contribution to the structural behaviour.

3 Conclusions

Two methods are used to analyse shell masonry elements for calculating the safety of the structure to ensure conservation, protect and safeguard of cultural heritage, to ensure the target 11.4 of the 11th goal of the 17 Sustainable Development Goals (SDGs).

The kinematic theorem of limit analysis is applied to a discrete model with simplex method of optimization that allows to calculate the collapse load and the mechanism for shell structures such as barrel vaults. It also allows to evaluate behaviour of a vault after an intervention of structural strengthening like reinforced with FRPM material. One advantage of the kinematic code is that only a few parameters are needed to obtain a response. Material properties considered as input parameter: (i) cohesion; (ii) friction angle; (iii) tensile strength. The innovation of this work with respect to the previous formulations of DLO are: (i) three-dimensional shell structure is studied; (ii) the effect of structural strengthening is considered. Different barrel vault shapes are considered: (i) round barrel vaults; (ii) ogive barrel vaults; (iii) lowered barrel vaults; (iv) skew barrel vaults. Dead and live loads in different positions are considered. Among the loads analysed: (i) uniformly distributed load; (ii) linear load; (iii) punctual force; (iv) inclined linear load. Regarding the boundary condition with four external boundaries, the available behaviour considered are: (i) rigid; (ii) symmetry plane; (iii) free.

In order to validate the formulation, the model is applied to experimental tests of vaults up to collapse. Close agreement with experimental examples Vermeltfoort [27] and homogenized limit analysis numerical solutions Milani [19] was obtained. Even in the case of reinforced vaults, a close agreement with Girardello [12] and Valluzzi [25, 26] was reached. Foreseen developments of the model are the design of optimal strengthening intervention for vaul ts, and the consideration of vertical movements at the supports of the vaults.

The static discrete algorithms for the membrane behaviour evaluate static regimes for masonry domes, of different typologies, considering the presence of a minimum tensile and shear strength before the structure begins to crack. Through the optimized solution, the static algorithms return the value of the forces acting on the structure, and the value of the multiplier for which the load for first cracking is reached. New possibilities arise to safeguard the artistic heritage. The first cracks do not lead to collapse, which to occur requires a load such as to trigger a mechanism, with an adequate number of cracks related to membrane forces and hinges related to the flexural regime.

The collapse, with load multiplier and collapse mechanism, is also considered with a model including flexural behaviour response. The analyses allow to consider the behaviour of a dome under different conditions, with and without cracks.

The fact that the algorithms consider the strength of the material allows us to comprehend the contribution of restoration interventions involving reinforcement on the dome. A typical example of reinforcement is represented by the traditional iron tie rods and by the more modern strips of composite material in which fibres with very high tensile strength are added to a mortar matrix that strengthen the structure. Moreover, the design of proper circumferential strengthening can be studied, to avoid cracking during intervention woks.

The model shows that with the same algorithm structure and some punctual modification different geometries can be analysed and compared. Considering single or double curvature domes, this feature modifies the response of the structure to loads and the positions of weight transfer. The role of geometry analysed shows the different response given by a double-curved dome, which distributes the load over the entire support surface, and a single-curved dome, which with to the development of shear forces concentrates the weight on the diagonal ribs.

The discrete algorithm allows: (i) to calculate the multiplier value, beyond which the structure cracks; (ii) to calculate collapse multiplier and mechanism in flexural regime; (ii) to consider different load conditions; (iv) to consider the performance of the structure in the presence of structural reinforcement; (v) to obtain, with limited input data, the forces acting on the boundaries of hyper-static structures, which often require very long and complex analyses; (vi) the discrete formulation allows to change the geometry of the structure quickly by the modification of only a few parameters, this allows to compare different geometries and configurations, thus considering the role of proportions in the design of ancient structures; (vii) the simple solution obtained can be used as benchmark for more complex analysis like Finite Element Model [2].

Amongst further future developments it will be possible to focus on non-axial symmetric conditions, such as effect of soil settlements or variable lateral loads, due to wind effect or seismic events.

Acknowledgements I would like sincerely to acknowledge my supervisor Professor Dario Coronelli and my co-supervisor Professor Giuliana Cardani for their help throughout this thesis and this research. I wish to extend my special thanks to Professor Matthew Gilbert from Sheffield University, Professor Thomas Boothby from Penn State University, and Professor Patrick Bamonte from Politecnico di Milano.

References

1. Agosta, L.: Comportamento strutturale delle cupole murarie, il caso di San Giorgio a Ragusa Ibla, Tesi Politecnico di Torino, supervisor: Tocci C., (2018)
2. Angjeliu, G., Coronelli, D., Cardani, G., Boothby, T.: Structural assessment of iron tie rods based on numerical modelling and experimental observations in Milan Cathedral. Eng. Struct. **206** (2020)
3. Bellin, I.F.: Le cupole di Borromini, la "scienza" costruttiva in età barocca, Mondadori Electa, Milano (2004)
4. Breymann, G.A.: Archi volte cupole (1885), Editrice Dedalo Roma (2003)
5. Boothby, T.E.: Analysis of masonry arches and vaults. Prog. Struct. Mat. Eng. **3**, 246–256 (2001)
6. Caffi, M.: Dell'abbazia di Chiaravalle in Lombardia. Illustrazione storico—Monumentale—Epigrafica, Milano (1842)

7. Cescatti, E., da Porto, F., Modena, C.: In-situ destructive testing of ancient strengthened masonry vaults. Int. J. Archit. Herit. (2018)
8. Como, M.: Statica delle costruzioni storiche in muratura: archi, volte, cupole, architetture monumentali, edifici sotto carichi verticali e sotto sisma, ARACNE editrice, Roma (2010)
9. D'Ayala, D.F., Tomasoni, E.: Three-dimensional analysis of masonry vaults using limit state analysis with finite friction. Int. J. Archit. Herit. (2011)
10. Flügge, W.: Stresses in Shells. Springer, Berlin (1960)
11. Gilbert, M., He, L., Smith, C.C., Le, C.V.: Automatic yield-line analysis of slabs using discontinuity layout optimization. Proc. R. Soc. A **470**, 20140071 (2014)
12. Girardello, P., da Porto, F., Modena, C., Valluzzi, M.R.: Comportamento sperimentale di volte in muratura rinforzate con materiali compositi a matrice inorganica, Anidis, Padova (2013)
13. Hawksbee, S., Smith, C.C., Gilbert, M.: Application of discontinuity layout optimization to three-dimensional plasticity problems. Proc. R. Soc. Math. Phys. Eng. Sci. **469**, 20130009 (2013)
14. Heyman, J.: Poleni's problem. Proc. Instn Civ. Engrs, Part 1, Struct. Eng. Group **84**, 737–759 (1988)
15. Heyman, J.: The Stone Skeleton, Structural Engineering of Masonry Architecture. Cambridge University Press, Cambridge (1995)
16. La Russa, M.F., et al.: Il Duomo di s. Giorgio a Ragusa Ibla: Individuazione dei materiali litici utilizzati, implicazioni architettoniche ed analisi delle forme di degrado., L'approccio multidisciplinare allo studio e alla valorizzazione dei beni culturali, Atti del workshop Siracusa (2005)
17. Lenzo, F.: Una cupola su colonne, Nuovi elementi per la comprensione di Sant'Agnese in Agone. In: Annali di Architettura, vol. 24, pp. 109–130 (2021)
18. Margani, G., et al.: Quattro studi sulla chiesa di San Nicolò l'Arena – Indagini storico-costruttive, Documenti e Quaderni del Dipartimento di Architettura e Urbanistica dell'Università di Catania, Catania (2004)
19. Milani, E., Milani, G., Tralli, A.: Limit analysis of masonry vaults by means of curved shell finite elements and homogenization. Int. J. Solids Struct. **45**, 5258–5288 (2008)
20. Poleni, G.: Memorie istoriche della gran cupola del Tempio Vaticano, Padova, Italy (1748)
21. Portioli, F., Casapulla, C., Gilbert, M., Cascini, L.: Limit analysis of 3D masonry block structures with non-associative frictional joint using cone programming. Comput. Struct. **143**, 108–121 (2013)
22. Smith, C.C., Gilbert, M.: Application of discontinuity layout optimization to plane plasticity problems. Proc. R. Soc. A **463**, 2461–2484 (2007)
23. Tomasoni, E.: Le volte in muratura negli edifici storici: tecniche costruttive e comportamento strutturale, Rome, Italy: Aracne Editrice (2008)
24. Ungewitter, G.: Lehrbuch der Gotischen Konstruktionen. 4th edn (neu bearbeitet von K. Mohrmann), vol. 2, Leipzig (1901)
25. Valluzzi, M.R., Valdemarca, M., Modena, C.: Behaviour of brick masonry vaults strengthened by FRP laminates. J. Compos. Constr. **5**(3), 163–169 (2001)
26. Valluzzi, M.R., Modena, C., de Felice, G., Current practice, and open issues in strengthening historical buildings with composites. Mater. Struct. **47**, 1971–1985 (2014)
27. Vermeltfoort, A.V.: Analysis and experiments of masonry arches. In: Lourenço, P.B, Roca, P. (eds) Proceedings of the Historical Constructions, Guimaraes PT (2001)

Displacement-Based Design of Geosynthetic-Reinforced Pile-Supported Embankments to Increase Sustainability

Viviana Mangraviti

Abstract Although the construction of concrete piles has a relevant environmental footprint, they are commonly used to reduce settlements of embankments on soft soil strata. A more sustainable choice to further reduce settlements (and, consequently, the number of piles) is to place geosynthetics below the embankment. However, existing design methods cannot calculate settlements at the embankment top and cannot be used to optimise the number of piles in a displacement-based design. In this note, an innovative model for assessing settlements at the top of Geosynthetic-Reinforced and Pile-Supported embankments induced by the embankment construction process is presented and validated against finite difference numerical analyses. The model is used to optimise the design of both piles and geosynthetic, and applied to a practical example, where the mass of CO_2 saved by designing geosynthetics to reduce the pile number.

Graphical Abstract

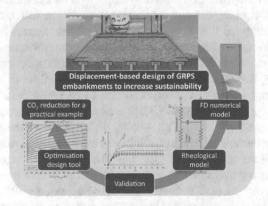

Keywords Geosynthetic-reinforced and Pile-Supported embankments · Displacement-based · SDG 9 · Sustainability · CO_2

V. Mangraviti (✉)
Chalmers University of Technology, 412 96 Gothenburg, Sweden
e-mail: viviana.mangraviti@chalmers.se

Politecnico di Milano, Piazza Leonardo da Vinci, 32, 20133 Milano, Italy

83

M. Antonelli and G. Della Vecchia (eds.), *Civil and Environmental Engineering for the Sustainable Development Goals*, PoliMI SpringerBriefs,
https://doi.org/10.1007/978-3-030-99593-5_7

1 Introduction

The expansion of urban areas featuring the last century resulted in the exploitation of increasingly large areas of territory, leading to deleterious impacts on the environment. Civil engineering is immensely contributing to the consumption of global energy reserves along with the severe exploitation of raw materials such as gravel, sand, and water [1]. Furthermore, the greenhouse gas emissions due to the production of concrete structures compared to both steel and wood structures is the highest [2]. For this reason, at least 2 of the 17 Sustainable Development Goals (SDGs) defined by United Nations (UN) in 2015 (i.e. both SDGs 9, "industry, innovation and infrastructures", and 11, "sustainable cities and communities", in Johnston [3]) can be linked to civil engineering.

In a context where all engineers can have a major influence towards a more sustainable development, geotechnical engineering has a crucial role in influencing the sustainability of a project. In fact, according to Abreu et al. [4], geotechnical engineering is one of the key contributing fields to sustainable development, since it faces a challenging dichotomy between delivering project goals (environmental, economic, and social) and maintaining sustainability. In fact, from a practical point of view, the exploitation of increasingly large areas of territory has led also to the construction of infrastructures under difficult geological and geotechnical conditions, requiring geotechnical engineers to find new and not always "environmentallly friendly" solutions. As an example, embankments for major infrastructures are more often realised in areas where soils are deformable, and to avoid unacceptable settlements, concrete piles are commonly employed as settlement reducers. Such "geo-structures", composed by embankment, foundation soil and concrete piles, are named Conventional Pile-Supported (CPS) embankments. The rigid inclusion induces the development of the "arching effect" within the embankment soil, reducing the portion of embankment load transferred to the soft soil, while stresses flow towards the piles, and consequently alleviating differential settlements. Depending on both the overall length of the infrastructure to be realised and the mechanical properties of the ground to improve, CPS embankments may require the installation of a huge number of concrete piles along different kilometres of infrastructure, leading to a huge outflow of both economic and environmental resources.

To further reduce settlements at the top (where infrastructures are placed) of CPS embankments, geosynthetic layers can be placed below the embankment. In the literature, Geosynthetic-Reinforced and Pile-Supported (GRPS) embankments were studied by several authors [5–11], and geosynthetics were found to effectively increase the transfer loads towards the piles, leading to several advantages: (i) a decreased number of inclusions (piles) needed; (ii) faster construction, and (iii) better control of differential settlements associated with soft soils. As a consequence, for equal admissible settlements at the embankment top, GRPS embankments need a fewer number of piles than CPS embankment, reducing the Embodied Carbon (EC, referring to carbon dioxide emitted during the manufacturing, transport, construction and the "end of life" of a material) due to the use of concrete.

Unfortunately, due to the lack of simplified methods, nowadays the design of GRPS embankments under a displacement-based perspective can only be done by using advanced numerical methods. In fact, the current design guidelines for GRPS embankments issued by several countries [12–15], adopt approaches based on the limit-equilibrium method that are not suitable neither to estimate settlements at the top of the embankment nor to ensure the serviceability of the geo-structure over its all lifetime [16]. In fact, these equilibrium arching models could possibly lead to an overestimation of the number of piles needed representing, from a sustainability perspective, a waste of both energy and resources. Nevertheless, to reach SDG 9, a more sustainable design perspective needs to be developed and spread in the next years.

With the aim of providing a more effective displacement-based design tool for GRPS embankments, Mangraviti et al. [11, 17] proposed an upscaled constitutive relationship to evaluate both differential and average settlements at the top of the central part of the embankment. The model was conceived as an extension of the one proposed by di Prisco et al. [18] for CPS embankments and derives from the interpretation of the results of a series of Finite Difference (FD) numerical analyses focusing on the construction process under drained conditions of the embankment.

In this chapter, the GRPS embankments capability to deform less than CPS embankments is firstly discussed by means of FD numerical results, expressed in terms of evolution of both average and differential settlements at the top of the embankment during construction. The upscaled constitutive relationship from Mangraviti et al. [17] is then shortly described and used as a tool to optimise the design of both piles and geosynthetic layer in a direct displacement-based perspective. Optimise the design means to employ the number of piles strictly necessary to undergo an admissible settlement at the top and, therefore, to achieve a more sustainable design, reducing EC.

The text is structured as it follows: in Sect. 1 the FD numerical model is presented and the results for both CPS and GPRS embankments are discussed. In Sect. 2 the constitutive relationship is briefly introduced and compared against numerical results. In Sect. 3 a non-dimensional chart to optimise the design of both piles and geosynthetics is provided and used to solve a practical example, where the EC for both CPS and GRPS embankments are calculated and compared.

2   Numerical Model

In the most general case, the design of both CPS and GRPS embankments is a three-dimensional problem. Nevertheless, when the embankment transversal width is significantly larger than its height, side effects may be disregarded and only one central axisymmetric cell can be considered as representative of the mechanical behaviour of the central part of system (Fig. 1a). The cell, whose diameter is equal to the spacing (s) between piles, includes: (*i*) the pile of length l and diameter d, (*ii*) the soft foundation soil, (*iii*) the embankment, whose height, h, evolves during

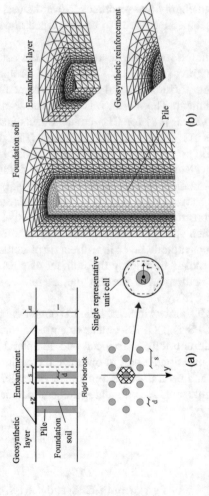

Fig. 1 **a** Problem geometry and representative axisymmetric cell; **b** numerical model

construction and (*iv*) the geosynthetic layer. The origin of both radial and vertical coordinates (r and z respectively) is in the centre of the pile top (Fig. 1a).

The problem has been numerically modelled by means of the finite difference numerical code FLAC3D [19]. Due to axisymmetry, only one quarter of the representative cell has been considered (Fig. 1b). The concrete end-bearing has been modelled as an elastic element. The mechanical behaviour of both the embankment and the foundation soil was modelled by means of an elastic-perfectly plastic constitutive relationship with a Mohr-Coulomb failure criterion and a non-associated flow rule. An elastic membrane element, characterised by axial stiffness J, has been used to model the reinforcement. When $J = 0$, the case of CPS embankment is obtained. Smooth interface elements were introduced between the pile and the foundation soil, whereas frictional interface elements were used between the geosynthetic and the soil. Normal displacements are not allowed on the lateral boundary and at the base of the domain.

To reproduce the construction process of the embankment, the numerical analysis has been subdivided in several stages. Each stage corresponds to the construction of 25 cm thick layer of the embankment under drained conditions. Therefore, the geometry of the spatial domain progressively evolves, adding at each stage a new layer of elements at the top of the model.

Even though a parametric study was conducted [11], for the sake of brevity, the results concerning only one reference geometry ($s = 1.5$ m, $d = 0.5$ m and $l = 5$ m) for different values of geosynthetic axial stiffness are hereafter illustrated in order to highlight the effectiveness of the geosynthetic layer in reducing settlements at the top of the embankment. The mechanical parameters used for the reference case are reported in Table 1. The dilatancy angle was found not to affect the mechanical processes of the system, although a slightly decrease in both average and differential settlements is observed within the embankment (the results are here omitted for the sake of brevity).

During the first step of construction, plastic strains develop in a narrow zone close to the top corner of the pile (defined as "process zone" in [18, 20], see Fig. 2). The evolution of this yielded zone is described by the process height h_p and when $h_p = h^*$ the plastic zone stops evolving.

According to di Prisco et al. [18], the mechanical response of the geo-structure can be described by using the following non-dimensional variables:

Table 1 Mechanical properties for the reference case

	Unit weight (kN/m^3)	Young's modulus (MPa)	Poisson's ratio (–)	Friction angle (°)	Dilatancy angle (°)
Foundation soil	18	1	0.3	30	0
Embankment	18	10	0.3	40	0
Pile	25	30000	0.3	–	–

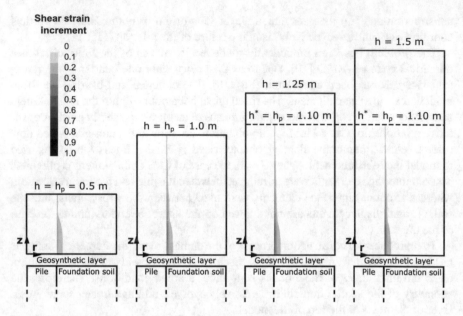

Fig. 2 Evolution of shear plastic strain during the construction process

$$H = h/d \tag{1}$$

$$U_{t,p} = \frac{u_{t,p}}{d}\frac{E_{\text{oed},f}/l}{\gamma} \quad \text{and} \quad U_{t,f} = \frac{u_{t,f}}{d}\frac{E_{\text{oed},f}/l}{\gamma} \tag{2}$$

$$U_{t,\text{diff}} = U_{t,f} - U_{t,p} \quad \text{and} \quad U_{t,\text{av}} = \frac{U_{t,f}(S^2 - 1) + U_{t,c}}{S^2} \tag{3}$$

where $u_{t,\blacksquare}$ are the average displacement at the top of the embankment, being the subscripts $\blacksquare = p$ and f when referring to pile (i.e. settlements and stresses for $0 < r < d/2$) and foundation soil ($d/2 < r < s/2$) respectively. $U_{t,\text{diff}}$ and $U_{t,\text{av}}$ are the non-dimensional differential and average settlements at the top; $S = s/d$ is the non-dimensional spacing; $E_{\text{oed},f}$ is the foundation soil oedometric modulus and the embankment unit weight.

The numerical results in Fig. 3 are plotted in non-dimensional planes where the system response was found to be uniquely defined if the non-dimensional geometrical ratios ($S = s/d$, $L = l/d$), the non-dimensional stiffness ratio ($E_{\text{oed},e}/E_{\text{oed},f}$, being $E_{\text{oed},e}$ the embankment soil oedometric modulus) and the embankment soil failure parameters (friction, ϕ'_e, and dilatancy angle, ψ_e, values) are kept constant. Due to the high difference in stiffness between piles and surrounding soil, differential settlements developing at the embankment base propagate to the top ($U_{t,\text{diff}} > 0$ in Fig. 3a). When H is sufficiently large ($H = H^* = h^*/d$, i.e. values highlighted with filled black rectangles in Fig. 3), differential settlements at the top of the embankment stop increasing, whereas average settlements continuously increase (Fig. 3b). As a consequence, any increase in load (e.g. the construction of the infrastructure above the embankment) for $H > H^*$ will not induce any increment of differential settlement in the transversal direction (r in Fig. 1), meaning that H^* in $r - z$ plane coincides with the height of the "plane of equal settlements", defined as the locus where the increment of differential displacements is negligible. On the contrary, $U_{t,av}$ continues to increase even for $H > H^*$, meaning that differential settlements in the longitudinal (y in Fig. 1a) direction increase. Therefore, H^* is a fundamental value to define the mechanical behaviour of the system when further loaded.

Due to the definition of non-dimensional variables (Eq. 3), the dashed lines inclined 1:1 in Fig. 3a,b represent the case of pile stiffness coincident with the foundation soil one and of $J = 0$ (i.e. nor piles neither geosynthetic are placed). The distance between $U_{t,av}$ and the 1:1 line is a measure of the effectiveness of both piles and geosynthetics as settlements reducers. The presence of the geosynthetic further reduce settlements with respect to the $J = 0$ case (Fig. 3a, b) meaning that, given a fixed value of settlement, less piles are needed when a geosynthetic layer with larger J is used. A direct connection between pile spacing (strictly related to the number of piles) and geosynthetic stiffness will be discussed in Sect. 3.

3 Mathematical Model

In di Prisco et al. and Mangraviti et al. [10, 11, 21] the critical interpretation of numerical results for GRPS embankments lead to the identification of 6 subdomains (Fig. 4a). All the subdomains are considered as elastic and behave in pseudo-

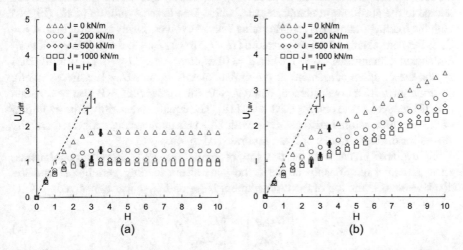

Fig. 3 Numerical results during embankment construction in terms of non-dimensional, **a** differential and **b** average settlements at the top of the embankment for different values of the geosynthetic axial stiffness

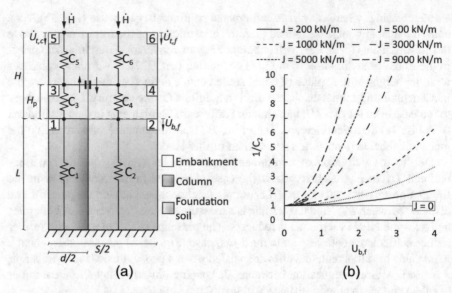

Fig. 4 a Rheological model for CPS embankments; **b** stiffness of reinforced foundation soil versus embankment base displacement for different J

oedometric conditions, whereas the arching effect [11, 17] is modelled as localised at the interface between subdomains 3 and 4. The mechanical behaviour of each subdomain can be reproduced by means of elastic springs and the arching effect is considered as localised and represented by a frictional slider placed in $z = H_p$ and $r = d/2$. In Fig. 4a, the non-dimensional compliances C_1 and C_2 represent the pile and the reinforced foundation soil respectively, whereas C_3 and C_4 are the compliances related to the portion of embankment H_p thick. Due to the evolution of H_p (Fig. 2) with the loading function $\dot{H}$, subdomains 3 and 4 evolves during construction. $C_5 = C_6$ is the compliance of the soil stratum $(H-H_p)$ thick. Due to the definition of non-dimensional quantities, $C_2 = 1$ when $J = 0$, whereas, when $J \neq 0$, C_2 is a function of non-dimensional settlement at the embankment bottom, $U_{b,f}$ (i.e. geosynthetic deformation). As a consequence, when $J = 0$, the model for GRPS embankments [11, 17] reduces to the one for CPS ones [18]. The equation describing the evolution of the non-dimensional stiffness $1/C_2$ with $U_{b,f}$ was numerically calibrated and the curves for different values of J are reported in Fig. 4b.

To evaluate settlements at the top of both CPS and GRPS embankments, an incremental relationship between the generalised loading variable $\dot{H}$ and the displacements at the top of the embankment $U_{t,\mathrm{diff}}$ and $U_{t,\mathrm{av}}$ was conceived:

$$\begin{bmatrix} \dot{U}_{t,\mathrm{diff}} \\ \dot{U}_{t,\mathrm{av}} \end{bmatrix} = \begin{bmatrix} C_{\mathrm{diff}} \\ C_{\mathrm{av}} \end{bmatrix} \dot{H} \tag{4}$$

C_{diff} and C_{av} represent the non-dimensional compliances of the overall system, analytically evaluated by (i) employing the rheological scheme illustrated in Fig. 4a; (ii) imposing the balance of momentum and compatibility conditions along the vertical direction; (iii) imposing the definition of plane of equal settlements ($\dot{U}_{t,\text{diff}} = 0$).

The obtained constitutive model in Eq. (4), describing the response of GRPS embankments depends on: (i) the mechanical parameters of piles, soil (Table 1) and geosynthetic (J); (ii) the geometrical variables (s, d, l) and (iii) the average ratio between horizontal and vertical stresses (k) in the portion of embankment in which irreversible strains accumulated. According to di Prisco et al. [18], k is the only parameter of the model which is not directly related on the system geometry/mechanical properties. k is not significantly affected by J [11] but only depends on the dilatancy angle.

A conservative value of the final height of the plane of equal settlements can be estimated as (di Prisco et al. [18]):

$$H^* = \frac{1}{2}\sqrt{\left[\frac{E_{\text{oed},e}}{E_{\text{oed},f}}\frac{L}{S^2}C_2\right]^2 + \frac{(S^2-1)}{\overline{k}\tan\phi'_{ss}}\left(\frac{E_{\text{oed},e}}{E_{\text{oed},f}}\frac{L}{S^2}\right)C_2} - \frac{1}{2}\left(\frac{E_{\text{oed},e}}{E_{\text{oed},f}}\frac{L}{S^2}\right)C_2 \quad (5)$$

As previously mentioned, for GRPS embankments C_2 in Eq. (5) is a function of $U_{b,f}$ and, according to the trend reported in Fig. 4b, the critical height H^* decreases for larger values of the geosynthetic axial stiffness J (i.e. H^* is maximum when $J = 0$). Eq. (5) will be used in the following section as a key ingredient to optimise the design of GRPS embankments.

The results obtained by integrating the mathematical model in Eq. (4) are compared with the numerical ones (see Fig. 3) in Fig. 5, where a good agree-

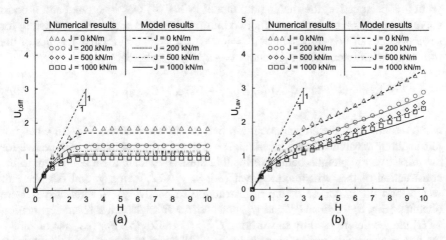

Fig. 5 Comparison between non-dimensional numerical results and the mathematical model in Eq. (4) in terms of **a** differential and **b** average settlements at the top of the embankment

ment is obtained for each value of J in terms of both differential (Fig. 5a) and average (Fig. 5b) settlement. The comparison between the mathematical model and the numerical results from a parametric study showed a satisfactory agreement [11], but the results are here omitted for brevity.

4   Optimisation of GRPS Embankments Design to Increase Sustainability and Practical Example

From a practical point of view, geotechnical engineers should design both piles (i.e. l, s and d) and geosynthetic layer (i.e. J) in order to have an average settlement at the top of GRPS embankments lower than an admissible settlement ($u_{t,av}^{amm}$). If end-bearing piles are considered, l is known, as well as the mechanical properties of the soft soil ($E_{oed,f}$).

Settlements accumulated during embankment construction are generally levelled thanks to the rollers compacting the soil. However, it is important that, after the construction of the infrastructure (i.e. either road superstructure or ballast plus rail track, where the rollers cannot be used anymore), the embankment does not settle more than expected. In fact, that would lead to possibly dangerous consequences to people and expensive damages to the infrastructure. As previously shown in Sects. 1 and 2, when $H > H^*$, differential settlements in r-direction stop increase (Fig. 5a), whereas average settlements (i.e. differential settlements in y-direction) continuously increase (Fig. 5b). As a consequence, to design GPRS embankments with a displacement-based approach, it is important: (i) that the final height of the embankment is larger than H^*, in order to avoid an increase in $U_{t,diff}$ and (ii) that the value of $U_{t,av}$ is equal or lower than the admissible one.

In this perspective, the rheological model in Eq. (4) was integrated and used to solve an optimisation problem: the maximum value of $S = s/d$ was evaluated, for several values of non-dimensional $J^* = (Jl)/(E_{oed,f} d^2)$ and for fixed values of the following non-dimensional efficiency: $1 - u_{t,av}^{amm}/u^*$, where:

$$u^* = \Delta q \left(l/E_{oed,f} + 0.5\Delta h/E_{oed,e} \right) \tag{6}$$

is the settlement at the embankment top induced by the load $\Delta q = \gamma \Delta h$ when nor piles neither geosynthetic are installed. Δq is the distributed load representative for the infrastructure weight, being Δh the thickness of embankment equivalent to the construction of the infrastructure layer ($\Delta h = \gamma_i \Delta h_i/\gamma$ being γ_i and Δh_i the unit weight and the thickness of the infrastructure respectively). Several curves were obtained (Fig. 6) for fixed ϕ'_e and ψ_e, and for $H > H^*$, where a safe side estimation of H^* can be determined by substituting $C_2^r = 1$ in Eq. (5). For the sake of safety, the curves were obtained by considering $u_{t,av}^{amm}$ equal to the maximum increment of average settlements at the top of the embankment (Eqs. 4, 3, 2) for $H > H^*$.

Fig. 6 Non-dimensional efficiency isolines for optimisation of GRPS embankments preliminary design (for $H > H^*$, $\phi'_e = 40°$ and $\psi_e = 0$)

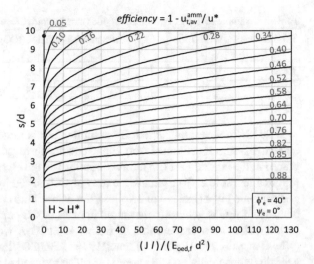

$1 - u_{t,av}^{amm}/u^*$ is a measure of the efficiency of the installation of both piles and geosynthetic: it will be equal to 0 when $u_{t,av}^{amm} = u^*$, whereas it is 1 for $u_{t,av}^{amm}/u^* \to 0$. As expected, by increasing the axial stiffness of the geosynthetic layer, a larger pile spacing can be chosen, possibly leading to a reduction of the concrete piles number and to a more sustainable design. However, it is worth noticing that, since the geosynthetic is more effective in transferring stresses to the piles when it deforms more (i.e. larger settlements), for large $u_{t,av}^{amm}/u^*$ values (i.e. low values of the efficiency in Fig. 6) the curves are more inclined and a small increase in J^* lead to choose piles with larger spacing (or smaller diameter). On the contrary, when very small settlements are admitted (i.e. larger efficiency is required), the increase in J^* induces small increase in s/d.

As an example, a GRPS embankment, with 25 m central part in r-direction, can be considered to be realised over a 5 m thick soft soil stratum (i.e. $l = 5$ m). The mechanical properties of the foundation soil, the concrete pile and the embankment soil are those reported in Table 1. A differential settlement of 6 mm at the top of the embankment is considered as admissible after a $\Delta h_i = 50$ cm thick superstructure is constructed. For the sake of simplicity, in this case, the unit weight of the infrastructure is (conservatively) assumed to be equal to the one of the embankment soil ($\gamma_i = \gamma \to \Delta q = \gamma_i \, \Delta h_i = 18 \cdot 0.5 = 9$ kPa).

As a first step, the isoline with *efficiency* $= 1 - (0.006)/[(9) \cdot (5/1346 + + 0.5 \cdot 0.5/13462)] = 0.82$ is individuated in Fig. 6. All the $s/d - Jl/(E_{oed,f} \, d^2)$ couples on the 0.82 isoline can be chosen to have a settlement at the top of the embankment equal to the admissible settlement (6 mm). In order to optimise both environmental and economic resources, the designer should look for the maximum values of both s and J that better suits the project needs.

To estimate the increment in sustainability (i.e. the reduction in CO_2 emissions) induced by considering a GRPS embankment instead of a CPS one, both $J = 0$ and J

Table 2 Example of design optimization: piles number and tCO$_2$ for CPS and GRPS embankment

Embankment	J (kN/m)	s (m)	Piles number	Piles (tCO$_2$)	Geogrid (tCO$_2$)	Total (tCO$_2$)
CPS	0	1.1	23	60.55	0	60.55
GRPS	1000	1.5	17	45.74	0.05	45.79

$= 1000$ kN/m were considered. By assuming a first tentative value of pile diameter equal to 0.5 m, s is estimated from Fig. 6 and the corresponding number of piles to be placed in the central part (25 m in r-direction) of the embankment are evaluated and reported in Table 2. To evaluate the tons of CO$_2$ saved by realizing 17 piles and the geosynthetic layer (instead of 23 piles), the EC for both concrete and geosynthetics is considered. EC is generally measured in mass of CO$_2$ emitted per mass of material. In particular, an average value of EC $= 1.08$tCO$_2$/t was chosen for the concrete (whose mass is 2500 kg/m^3) [22] and EC $= 2.36$tCO$_2$/t was considered for a woven geogrid (whose mass is 0.53 kg/m^2) [23]. The reduced number of piles employed for the GRPS embankment led to a 24% reduction of CO$_2$ emissions if compared to the CPS embankment. The same calculation may be repeated for a different value of d, in other to further optimise the sustainability of the project.

The charts in Fig. 6 represent a very effective and quick tool to design GRPS embankments in a displacement-based perspective and to optimise the number of piles for a more sustainable design. Several design charts were obtained for different ϕ'_e and ψ_e and are here omitted for the sake of brevity.

5 Conclusion

In this chapter a mathematical model capable of reproducing the mechanical response of GRPS embankments is presented. The model considers the embankment height as a generalised loading variable and allows to evaluate settlements at the top of the embankment during construction under drained conditions. The model is highly innovative since it represents an effective tool during the preliminary design of GRPS embankments, including a displacement-based approach. In order to move a step forward towards a more sustainable construction of infrastructures (SDG 9) and reduce as much as possible the CO$_2$ emissions due to the construction of unnecessary concrete piles, one chart was provided to optimise the design of both piles and geosynthetic in a displacement-based perspective. The calculation of CO$_2$ saved for the practical example considered confirmed that GRPS embankments are a more sustainable choice if compared to CPS embankments.

Acknowledgements All the numerical results have been obtained by using the commercial code FLAC3D within the framework of the Itasca Education Partnership (IEP) program. I am grateful to Itasca Consulting Group and Harpaceas for the use of the software license. The financial support from Nordforsk (project #98335 NordicLink) is greatly appreciated. I would like also to acknowledge

Prof. Claudio di Prisco, Prof. Jelke Dijkstra and PhD Luca Flessati, for supporting me during this research.

References

1. Dixit, M.K., Fernández-Solís, J.L., Lavy, S., Culp, C.H.: Identification of parameters for embodied energy measurement: a literature review. Energy Build. **42**(8), 1238–1247 (2010). https://doi.org/10.1016/j.enbuild.2010.02.016
2. Latawiec, R., Woyciechowski, P., Kowalski, K.J.: Sustainable concrete performance—CO_2-emission. Environ. MDPI **5**(2), 1–14 (2018). https://doi.org/10.3390/environments5020027
3. Johnston, R.B.: Arsenic and the 2030 Agenda for sustainable development. In: Arsenic Research and Global Sustainability—Proceedings of the 6th International Congress on Arsenic in the Environment, AS 2016, pp. 12–14 (2016). https://doi.org/10.1201/b20466-7
4. Abreu, D.G., Jefferson, I., Braithwaite, P.A., Chapman, D.N.: Why is sustainability important in geotechnical engineering?, pp. 821–828 (2008). https://doi.org/10.1061/40971(310)102
5. Han, J., Gabr, M.A.: Numerical analysis of geosynthetic-reinforced and pile-supported earth platforms over soft soil. J. Geotech. Geoenviron. Eng. **128**(1), 44–53 (2002). https://doi.org/10.1061/(asce)1090-0241(2002)128:1(44)
6. Stewart, M.E., Filz, G.M.: Influence of clay compressibility on geosynthetic loads in bridging layers for column-supported embankments, pp. 1–14 (2005). https://doi.org/10.1061/40777(156)8
7. Yan, L., Yang, J.S., Han, J.: Parametric study on geosynthetic-reinforced pile-supported embankments. Adv. Earth Struct. 255–261 (2006). https://doi.org/10.1061/40863(195)28
8. Liu, H.L., Ng, C.W.W., Fei, K.: Performance of a geogrid-reinforced and pile-supported highway embankment over soft clay: case study. J. Geotech. Geoenviron. Eng. **133**(12), 1483–1493 (2007). https://doi.org/10.1061/(ascc)1090-0241(2007)133:12(1483)
9. Wijerathna, M., Liyanapathirana, D.S.: Load transfer mechanism in geosynthetic reinforced column-supported embankments. Geosynth. Int. **27**(3), 236–248 (2020). https://doi.org/10.1680/jgein.19.00022
10. di Prisco, C., Flessati, L., Galli, A., Mangraviti, V.: A Simplified approach for the estimation of settlements on piled foundations. Lect. Notes Civil Eng. **40**, 640–648 (2020). https://doi.org/10.1007/978-3-030-21359-6_68
11. Mangraviti, V.: Theoretical modelling of embankments based on piled foundations. Ph.D. thesis, Politecnico di Milano (2021)
12. BSi.: BS 8006:1995—Code of practice for strengthened/reinforced soils and other fills Amd 1. British Standards Institution. ISBN 978-0-580-53842-1 (1995)
13. EBGEO.: Empfehlungen für den Entwurf und die Berechnung von Erdkörpern mit Bewehrungen aus Geokunststoffen—EBGEO (2010)
14. ASIRI.: Recommandations pour le dimensionnement, l'execution et le controle de l'amelioration des sols de fondation par inclusions rigides. ISBN 978-2-85978-462-1 (2012)
15. CUR226.: Design Guideline Basal Reinforced Piled Embankments. CRC Press (2016)
16. King, D.J., Bouazza, A., Gniel, J.R., Rowe, R.K., Bui, H.H.: Serviceability design for geosynthetic reinforced column supported embankments. Geotext. Geomembr. **45**(4), 261–279 (2017). https://doi.org/10.1016/j.geotexmem.2017.02.006
17. Mangraviti, V., Flessati, L., di Prisco, C.: Mathematical modelling of the mechanical response of Geosynthetic-Reinforced and Pile-Supported embankments. Under Rev. (2022)
18. di Prisco, C., Flessati, L., Frigerio, G., Galli, A.: Mathematical modelling of the mechanical response of earth embankments on piled foundations. Geotechnique **70**(9), 755–773 (2020). https://doi.org/10.1680/jgeot.18.P.127

19. Itasca.: FLAC3D v.5.0—Fast Lagrangian analysis of continua in three dimensions. User manual. Itasca Consulting Group, Minneapolis. Minneapolis (2012)
20. Flessati, L., di Prisco, C., Corigliano, M., Mangraviti, V.: A simplified approach to estimate settlements of earth embankments on piled foundations: the role of pile shaft roughness. Eur. J. Environ. Civil Eng. (2022). https://doi.org/10.1080/19648189.2022.2035259
21. Mangraviti, V., Flessati, L., di Prisco, C.: Geosynthetic-reinforced and pile-supported embankments: theoretical discussion of finite difference numerical analyses results. Under Rev. (2022)
22. Koerner, G.R.: Relative sustainability (i.e., Embodied Carbon) calculations with respect to applications using traditional materials versus geosynthetics. Geosynthetic Institute (2019)
23. Raja, J., Dixon, N., Fowmes, G., Frost, M., Assinder, P.: Obtaining reliable embodied carbon values for geosynthetics. Geosynth. Int. **22**(5), 393–401 (2015). https://doi.org/10.1680/jgein.15.00020

Characterization and Monitoring of an Unstable Rock Face by Microseismic Methods

Zhiyong Zhang ⓘ

Abstract Unstable rock slopes are likely to cause rockfalls, threatening human lives and properties, industrial activities, and transportation infrastructures in mountain areas. There is an increasing demand to forecast and mitigate the potential damage of rockfalls by developing a reliable early warning system. In this thesis, an unstable mountain slope in northern Italy was selected as the research target. A microseismic monitoring network has been operating since 2013 as a field research laboratory to study the microseismic monitoring technique in the perspective of developing rockfall early warning systems. Locating microseismic events is a basic step of this technique to obtain the location of developing cracks as possible precursors of rockfalls. However, it is still a challenging task due to the heterogeneity of fractured rock slopes. The main purpose of this thesis is to address the issues related to event localization for microseismic monitoring strategy applied to the unstable rock face.

Graphical Abstract

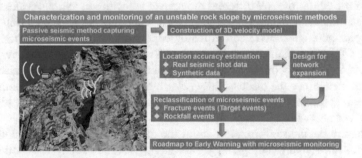

Keywords Landslide disaster · Rockfall · SDG 11 · Microseismic monitoring · Early warning system · Event localization

Z. Zhang (✉)
Department of Civil and Environmental Engineering, Politecnico di Milano, Piazza Leonardo da Vinci 32, 20133 Milano, Italy
e-mail: zhiyong.zhang@polimi.it

1   Introduction

Extension of human activities in mountain areas has given rise to worries about unstable rock slopes. Unstable rock slopes are likely to result in rockslides, causing serious threats to human safety and properties, industrial activities, and transportation infrastructures. Therefore, it is an increasingly significant issue to prevent and forecast dangerous incidents caused by unstable rock slopes. Management of rockslide hazards requires adequate information to better understand the mechanisms of rock slope failure. Monitoring an unstable rock slope as an early warning system to identify the danger of mass movement is the first step to assess the risk of rockslides. In this perspective, building an early warning system for rockslide hazards is consistent with the 11th sustainable development goals (SDGs) set by the United Nations, which is 'Goal 11: Make cities inclusive, safe, resilient and sustainable'. More specifically, research in this field exclusively serves for one specific objective (11.5) under Goal 11: 'By 2030, significantly reduce the number of deaths and the number of people affected and substantially decrease the direct economic losses relative to global gross domestic product caused by disasters, including water-related disasters, with a focus on protecting the poor and people in vulnerable situations.' In details, the early warning system plays an important role to mitigate or avoid the deaths and economic losses caused by rockslide-related disasters in mountain areas throughout the world.

Rockslides, or more specifically 'rock' landslides, are a group of landslides composed of rocks rather than soils. According to the commonly used classification of landslide types [1], rockslides can be divided into different subclasses based on the type of movements (fall, topple, slide, spread, flow and slope deformation). Different rockslides present different movement velocities at various scales and the mechanical behavior of the rockslides is best related to the rock material. Due to the complex geological and geomorphological conditions on unstable rock slopes, rock mass failure contains more than one class of rockslides. For fractured brittle rock mass, the stability of the rock slope is primarily controlled by the existence of discontinuities. Since the discontinuity pattern of rock mass distributes both on the surface and in the subsurface, the designed monitoring system should investigate on the rock face and within the rock mass.

To acquire detailed information about the distribution of brittle discontinuities within rock mass, many geological, geomorphological and geophysical methods can be adapted. Remote-sensing and aerial techniques are usually employed to map the surface area of the unstable rock slope to indicate the topography. However, mapping investigation techniques limit the observation on the rock surface. Geotechnical techniques mainly including boreholes and trenching can provide detailed geological information, indicating the vertical boundary of the slide and the properties necessary for the slope stability analysis. However, these techniques can only monitor a few critical points or lines and the difficulty of drilling the unstable and steep slopes usually limits their application [2]. On the other hand, geophysical methods (e.g., seismic, electrical, electromagnetic and gravimetric methods) can detect the petrophysical parameters under the rock surface. Changes of the geophysical parameters characterizing the rock mass are likely to be generated by the rock mass modifications due to its instability.

Geophysical methods that are commonly used to investigate landslides are summarized in several review papers [3–6]. For rockslides studies, passive seismic techniques are the most frequently used method. The reason may be that these techniques can be carried out to characterize and monitor the stability of the rock slope. The application fields of passive seismic techniques mainly include two areas. The first application field is seismic noise measurements which can analyze the seismic response of rock slopes by processing seismic noise that is not related to incipient fracturing (e.g. earthquakes, ambient seismic vibrations). The second application field is to detect and locate microseismic events generated by fracture propagation inside unstable rock masses.

The capability of passive seismic methods in detecting and locating microseismic events makes it an efficient option to be integrated into early warning alarm systems. By locating microseismic events generated within an unstable rock mass, we can locate the growing cracks and understand the slide kinematics and triggering mechanisms of future collapses [7]. Therefore, microseismic events can be the precursors before the occurrence of a macroscopic failure of rockfall or rockslide.

Concerning the applications of passive methods in rockslides, several studies were devoted to analyzing microseismic signals as possible precursors of rockfalls [8–11]. These studies show the usefulness of a microseismic monitoring network in the early warning alarm systems for rockfall risk on unstable rock slopes. For early warning purposes, more applications of this method are used in different sites [12–17]. These studies showed that the larger amounts of recorded seismic signals require a classification procedure to distinguish microseismic events from the noise. More importantly, these classified microseismic events also need to be located, because the localization of microseismic events is necessary to understand which part of the rock mass is cracking in order to be able to recognize the rockfall risk. Location results will also help to select the microseismic events close to the target monitored area and exclude small rockfalls and the events that are far from the monitored area, generated by other sources.

Although hypocenter localization of seismic events (usually followed by data acquisition and event classification) has been a basic step in microseismic monitoring, it is still a challenging task. One main reason that causes difficulties for a reliable event location is that the seismic velocity distribution is highly heterogeneous within unstable slopes so that the waveforms experience strong scattering and high attenuation. Therefore, contrary to what is usually adopted in large scale studies for earthquake localization, it is impossible to use a homogeneous or layered model to characterize the fractured zone in the limited rock volumes monitored by small-scale microseismic networks. Moreover, short distances between sources and receivers make it difficult to distinguish P and S waves, and thus direct estimations of source distance are impeded.

In this thesis, the localization problem was addressed for a microseismic monitoring network installed on the unstable rock cliff of Mount San Martino for early warning purposes. Based on the event location, the classification of microseismic events was revisited to extract fracture events. Finally, a procedure for the design of an expanded network was proposed to improve location accuracy.

2 Methods and Results

The research target is the limestone rock face of Mount San Martino that is threatening the town of Lecco (northern Italy) due to its historical rockfall events (Fig. 1). A microseismic monitoring system was installed on the mountain for early warning purposes in 2013 and has continuously recorded the data since then.

Fig. 1 The study site in the southern part of Mount San Martino. **a** Front view of Mount San Martino from Lecco. The yellow rectangle indicates the monitored area where the monitoring network was deployed. **b** Rock face with the positions of the five three-component geophones and the acquisition board of the microseismic monitoring system. The red arrow indicates the partially detached pillar below geophone 4. **c** Photogrammetric model of the monitored area with the orientations of the five geophones

To construct the three-dimensional (3D) velocity model required for reliable hypocenter localization, a seismic tomographic survey was performed on the mountain surface above the rock face. Source test measurements were conducted to select a suitable source capable of triggering all geophones. Hammer emerged among the other tested options as the preferred source because it performed like the seismic gun in terms of signal energy and spectral content while demonstrating superiority in terms of portability and flexibility in the harsh environment of the study site. We used GeoTomCG software [18] which performs three-dimensional tomographic inversion. The results of seismic traveltime inversion showed high heterogeneity of the rock mass with significant contrasts in velocity distribution (Fig. 2). Low velocities were found at the shallow depth on top of the rock cliff and intermediate velocities were observed in the most critical area of the rock face corresponding to a partially detached pillar. Two sensitivity tests were implemented to evaluate the resolution and stability of the inversion [19].

After obtaining the 3D velocity model from the inversion, the global grid search location method [20] was selected for event localization and the misfit function was defined based on the Equal Differential Time (EDT) method [21]. Seismic shots with known positions were located to estimate the location accuracy on the upper part of the rock mass. The hypocenter misfits were around 15 m with the 5 geophones of the microseismic network and the error was significantly decreased compared to the results produced by a constant velocity model. Besides, the analysis of location accuracy was extended to the whole volume of the rock mass by using synthetic traveltimes affected by random errors consistent with the noise level observed on real data (Fig. 3). Although three geophones of the seismic network are installed

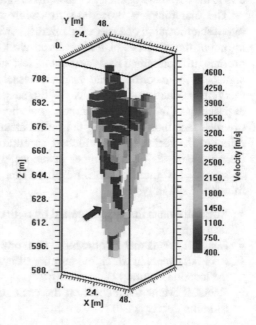

Fig. 2 Velocity model obtained from 3D tomographic inversion using a 2 × 2 × 4 m grid. Colored blocks indicate the velocity field in the rock mass where at least one ray travels. The red arrow indicates the partially detached pillar on top of which geophone 4 is installed

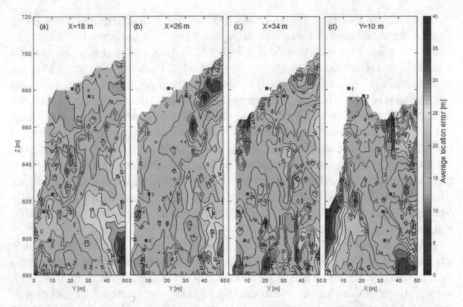

Fig. 3 Average location error on vertical slices cutting the 3D model at X = 18, 26, 34 m (**a, b, c**) and at Y = 10 m. **d** Red squares indicate the positions of the five geophones projected on each slice

near the critical area where the partially detached rock pillar was located, accuracy in this area is not better than the accuracy on the upper part of the rock face probably due to the high heterogeneity of rock mass and the higher velocities in this area [22].

The classification procedure for microseismic events was updated by the hypocenter location of a subset of microseismic events with high data quality, which is one of the novelties of this research work [23]. The microseismic events were preliminarily classified into two subclasses: suspected rockfall events with multiple signals in the recordings and suspected fracture events with a single signal in the recordings. The location results for the suspected rockfall events almost met our expectations (Table 1). Most of the signals in rockfall events were located on or near the rock face thus confirming the initial classification. The last column in Table 1 refers to the rockfall type, which is an additional information that we can derive from the analysis of the number of recorded signals combined with their time of occurrence and their location. Based on that, rockfall events can be sub-classified into four rockfall types:

- Rockfall events that are generated by only one falling stone bouncing on the rock face (Single);
- Rockfall events that involve more than one stone (Multiple);
- Rockfall events in which the number of involved stones is uncertain considering the location accuracy (Uncertain);
- Rockfall events that occurr on the rock surface at the summit of the rock face (Summit debris noise).

Table 1 Summary of the location results for 10 rockfall events. "N_{pick}" refers to the number of signals that can be picked from all the five geophones in each rockfall event. "N_{face}" refers to the number of signals that were located on or near the rock face. "N_{far}" refers to the number of signals that were located on the boundary of the model and far from the rock face. "N_{unc}" refers to the number of signals that were unexpectedly located in the middle of the model rather than on the rock face so that the localization result is considered uncertain. "Rockfall type" is explained in the text

Rockfall event	N_{pick}	N_{face}	N_{far}	N_{unc}	Rockfall type
1	4	4			Single
2	2	2			Multiple
3	5	5			Multiple
4	3	2	1		Multiple
5	3	3			Uncertain
6	1	1			Summit debris noise
7	3	2	1		Multiple
8	7	3	3	1	Multiple
9	2	2			Multiple
10	3	3			Single

For the suspected fracture events, Table 2 summarizes the location results for 20 selected fracture events. Based on the hypocenter locations, these fracture events could be reclassified into five groups: events that are located inside the rock mass (Fracture); events that are located on the rock surface at the summit of the rock face and are probably related to the movement of the debris covering the summit (Rockfall); events that are located on the vertical boundaries of the model (Far event); events that are located near the rock face so that they are likely generated by a fracture although we cannot exclude a probable single stone rockfall hitting the rock face only once and then disappearing without generating any other signal from a lower point below the monitored area (Suspected fracture); and events that are located at the lower boundary of the model near the rock face so that we suspect that they are fracture events occurring within the lower section of the unstable area of the rock mass, i.e., the lower part of the unstable rock pillar, although we cannot exclude the case of a single stone hitting the rock face only once at a point located below the

Table 2 Summary of the location results for 20 fracture events

Location of fracture events	Reclassification	Number of events
Inside rock mass	Fracture	4
Rock surface	Rockfall	4
Vertical boundary of the model	Far event	2
Near rock face	Suspected fracture	8
Lower boundary of the model	Suspected fracture (outside the area)	2

monitored area (Suspected fracture (outside the area)). Based on the results, only a few (4 out of 20) events were located inside the rock mass at such a distance from the rock face that they cannot be confused with rockfalls. Further improvements in location accuracy are necessary to distinguish suspected fracture events that were located close to the rock face from rockfalls. On the other hand, the hypocenter location helped to identify events generated outside the monitored area and rockfalls that were located on the rock surface at the summit of the rock face. This feasibility study shows that the hypocenter location is a promising method to improve the final classification of microseismic events.

Another original contribution of the thesis is to suggest a procedure for designing an expanded network in order to improve the localization accuracy in the most critical part of the rock mass [24]. Additional geophones were progressively added to the existing network, selecting their positions with special care to improve the azimuth coverage and source-receiver distance coverage. Three progressively expanded networks with 9, 13 and 15 geophones were studied to explore the improvement in location accuracy and it was observed that the location error was reduced from 12–24 m (5 geophones) to 4–6 m (15 geophones) (Fig. 4). The relatively low location errors in the critical part with 15 geophones would help to improve the classification of microseismic events. The method can be used in near future for expanding the

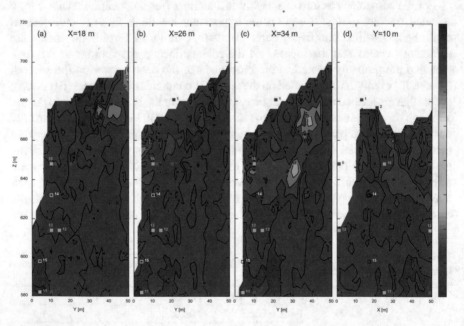

Fig. 4 Average location error on the vertical slices at X = 18 m, **a** 26 m, **b** 34 m, **c** and the vertical slice at Y = 10 m **d** with a network of 15 geophones. Red squares and other colored squares indicate respectively the positions of the original five geophones and additional geophones projected on each slice (Colorbar: 0–40 m as in Fig. 3)

network transforming the current network with 15 channels using 5 three-component (3C) geophones into a network of 15 single-component (1C) geophones.

Summarizing, the main novelties of this research work are:

- the design and the practical solutions adopted to perform the tomographic experiment in a harsh environment to derive a reliable 3D velocity model;
- the analysis of the localization accuracy extended to the whole monitored rock mass;
- the reclassification of microseismic events based on hypocenter location;
- the methodology proposed to design an expanded network to improve the localization and classification of microseismic events.

3 Conclusions

Microseismic monitoring, as one application of passive seismic techniques, has been increasingly used in the studies of unstable rock slopes to develop an early warning alarm system. This system would work for decreasing the damages caused by rockslide-related disasters, serving for the Goal 11 of SDGs. In order to recognize the areas with potential rockfall or rockslide risk triggered by unstable rock slopes, localization of seismic events is required to trace the propagation of fractures within the rock mass.

The first two goals of this thesis were to construct a 3D velocity model that is required for reliable hypocenter localization and to estimate the location accuracy based on the obtained 3D velocity model. To construct the 3D velocity model for the study area, a tomographic survey was carried out after performing source tests to select a suitable source. Then seismic traveltime inversion was implemented by using first-arrival picks to obtain the velocity distribution within the rock mass. Two sensitivity tests were performed to evaluate the resolution and stability of the inversion. To estimate the location accuracy, the global grid search method was first selected for event localization and the misfit function was defined based on the EDT method. Then, the location accuracy was estimated in two steps: 1-using the data of seismic shots with known positions (performed in the tomographic survey) to estimate the location accuracy in the upper part of the rock face; 2-assessing the expected location accuracy for the whole rock mass, particularly in the critical area, by using synthetic data affected by random errors.

However, the limitations of the present work must be noted for future improvements. Due to the harsh investigated area, the layout of the source positions for the tomographic survey was constrained on the rock surface at the summit of the rock face. This causes the inverted velocity model to only cover a limited volume of the rock mass, especially in the lower part of the monitoring area. Therefore, the localization results should be further improved, for example, by better extrapolating the velocity field to the neighboring areas of the current 3D model and by testing new misfit functions, also including a travel-time quality factor. Moreover, in order to enlarge the raypath coverage within the rock mass, some source positions can be designed on the rock face (if possible) for future tomographic investigation.

Although the location accuracy is not small enough to distinguish rockfalls from rock cracks occurring near the exposed face of the rock mass, the location algorithm can be used to improve the classification procedure by distinguishing the events occurring in the monitored area from rockfalls or other events occurring far from the monitored area. Moreover, the location accuracy can be improved by introducing more geophones to the current 5-geophone network. From these points of view, the location results implicate that the location algorithm based on the 3D velocity model can be applied to realize two other objectives of this thesis, which represent two original contributions of this research work: to explore the usefulness of hypocenter location in the classification of microseismic events and to design an expanded network to improve the location accuracy.

The results obtained in this thesis show the necessity of expanding the geophone network to improve the location accuracy and then improve the classification of the microseismic events. If event localization is accurate enough, its results can be used to distinguish rockfall events from fracture events because rockfalls only occur on the rock surface. Another future work would be exploring some efficient denoising methods to process the microseismic events with relatively lower SNRs to get reasonable arrival times for event localization. After this procedure, more fracture events can be extracted by the classification based on hypocenter location. This group of fracture events will be used to analyze the evolution of microseismic activities with time and space to indicate the most unstable area on the rock face. Moreover, the evolution of microseismic activities can be correlated with meteorological parameters to show the effects of temperature and rainfall on triggering the propagation of fractures probably leading to future collapse.

The take-home messages can be summarized as follows:

- microseismic events related to fracturing propagation in rock mass (target events), as the precursor information of rockslide hazards, can be recorded by using a proper microseismic monitoring system;
- the target events can be recognized in all the recordings by a reclassification procedure based on hypocenter location;
- the optimal monitoring network can be designed following simple guidelines to achieve a high location accuracy;
- the increasing risk of rockslides can be predicted on the basis of accurate spatial and temporal distribution of target events.

The perspective of the whole research is the implementation of an early warning methodology applicable to unstable rock slopes. Although this goal has not been achieved yet, the results of this work contribute to reduce the distance to the final objective and will provide valid support for scientists, technicians and decision makers to face future sustainability challenges related to rockslide hazards. Moreover, this work also contributes to fulfil the Goal 11 of SDGs, which is, in short, significantly reduce the loss in lives and property caused by disasters, including rockslide-related disasters mentioned in this research, with a focus on protecting the poor and people in vulnerable situations by 2030.

Acknowledgements I am grateful to my three supervisors Prof. Luigi Zanzi, Dr. Diego Arosio, Dr. Azadeh Hojat for their help throughout this thesis. This research was funded by China Scholarship Council (CSC), grant number 201606420051, to support me during my research activities in Italy.

References

1. Hungr, O., Leroueil, S., Picarelli, L.: The Varnes classification of landslide types, an update. Landslides **11**, 167–194 (2014)
2. Jongmans, D., Garambois, S.: Geophysical investigation of landslides: a review. Bull. Société Géologique Fr. **178**(2), 101–112 (2007)
3. Bogoslovsky, V.A., Ogtlvy, A.A.: Geophysical methods for the investigation of landslides. Geophysics **42**, 562–571 (1977)
4. McCann, D.M., Forster, A.: Reconnaissance geophysical methods in landslide investigations. Eng. Geol. **29**, 59–78 (1990)
5. Hack, R.: Geophysics for slope stability. Surv. Geophys. **21**, 423–448 (2000)
6. Pazzi, V., Morelli, S., Fanti, R.: A review of the advantages and limitations of geophysical investigations in landslide studies. Int. J. Geophys. (2019)
7. Arosio, D., Longoni, L., Papini, M., Scaioni, M., Zanzi, L., Alba, M.: Towards rockfall forecasting through observing deformations and listening to microseismic emissions. Nat. Hazards Earth Syst. Sci. **9**, 1119–1131 (2009)
8. Amitrano, D., Grasso, J.R., Senfaute, G.: Seismic precursory patterns before a cliff collapse and critical point phenomena. Geophys. Res. Lett. **32**, 1–5 (2005)
9. Senfaute, G., Duperret, A., Lawrence, J.A.: Micro-seismic precursory cracks prior to rock-fall on coastal chalk cliffs: a case study at Mesnil-Val, Normandie, NW France. Nat. Hazards Earth Syst. Sci. **9**, 1625–1641 (2009)
10. Walter, M., Schwaderer, U., Joswig, M.: Seismic monitoring of precursory fracture signals from a destructive rockfall in the Vorarlberg Alps. Austria. Nat. Hazards Earth Syst. Sci. **12**, 3545–3555 (2012)
11. Arosio, D., Longoni, L., Papini, M., Zanzi, L.: Analysis of microseismic activity within unstable rock slopes. In: Scaioni, M. (ed) Modern Technologies for Landslide Monitoring and Prediction, pp. 141–154. Springer, Berlin, Heidelberg (2015)
12. Spillmann, T., Maurer, H., Green, A.G., Heincke, B., Willenberg, H., Husen, S.: Microseismic investigation of an unstable mountain slope in the Swiss Alps. J. Geophys. Res. Solid Earth **112**, 1–25 (2007)
13. Blikra, L.H.: The Åknes rockslide; monitoring, threshold values and early-warning. In: Landslides and Engineered Slopes. From the Past to the Future. Proceedings of the 10th International Symposium on Landslides and Engineered Slopes, p. 2124 (2008)
14. Amitrano, D., Arattano, M., Chiarle, M., Mortara, G., Occhiena, C., Pirulli, M., Scavia, C.: Microseismic activity analysis for the study of the rupture mechanisms in unstable rock masses. Nat. Hazards Earth Syst. Sci. **10**, 831–841 (2010)
15. Helmstetter, A., Garambois, S.: Seismic monitoring of schilienne rockslide (French Alps): analysis of seismic signals and their correlation with rainfalls. J. Geophys. Res. Earth Surf. **115**, 1–15 (2010)
16. Codeglia, D., Dixon, N., Fowmes, G.J., Marcato, G.: Strategies for rock slope failure early warning using acoustic emission monitoring. In: IOP Conference Series: earth and Environmental Science, vol. 26 (2015)

17. Colombero, C., Comina, C., Vinciguerra, S., Benson, P.M.: Microseismicity of an unstable rock mass: from field monitoring to laboratory testing. J. Geophys. Res. Solid Earth **123**, 1673–1693 (2018)

18. Tweeton, D.R.: GeoTomCG, Three dimensional geophysical tomography software, Apple Valley, Minnesota: GeoTom LLC (2012)

19. Zhao, D., Hasegawa, A., Horiuchi, S.: Tomographic imaging of P and S wave velocity structure. J. Geophys. Res. **97**, 19909–19928 (1992)

20. Lomax, A., Michelini, A., Curtis, A.: Earthquake location, direct, global-search methods. In: Meyers, R.A. (ed) Encyclopedia of Complexity and Systems Science, pp. 2449–2473, Springer (2009)

21. Font, Y., Kao, H., Lallemand, S., Liu, C.S., Chiao, L.Y.: Hypocentre determination offshore of eastern Taiwan using the maximum intersection method. Geophys. J. Int. **158**, 655–675 (2004)

22. Zhang, Z., Arosio, D., Hojat, A., Zanzi, L.: Tomographic experiments for defining the 3D velocity model of an unstable rock slope to support microseismic event interpretation. Geoscience **10**, 327 (2020)

23. Zhang, Z., Arosio, D., Hojat, A., Zanzi, L.: Reclassification of microseismic events through hypocenter location : case study on an unstable rock face in Northern Italy. Geoscience **11**, 37 (2021)

24. Zhang, Z., Arosio, D., Hojat, A., Zanzi, L.: Optimal design for expanding a microseismic monitoring network on an unstable rock face in Northern Italy. In: EAGE 4rd Asia Pacific Meeting Near Surface Geoscience & Engineering. Ho Chi Minh, Vietnam (2021)

Printed in the United States
by Baker & Taylor Publisher Services